GOVERNORS STATE UNIVERSITY LIBRARY
P9-BYH-142
3 1611 00193 9484

The Nuclear Power Debate

Desaix Myers III

Published in cooperation with the Investor
Responsibility Research Center, Washington, D.C.

The Nuclear Power Debate
Moral, Economic, Technical, and Political Issues

PRAEGER PUBLISHERS
Praeger Special Studies

New York • London • Sydney • Toronto

Library of Congress Cataloging in Publication Data

Myers, Desaix B
The nuclear power debate.

(Praeger special studies in U.S. economic, social, and political issues)
Includes bibliographical references.
1. Atomic power industry—United States. 2. Atomic power—Security measures. I. Title.
HD9698.U52M93 1977 338.4'7'621480973 75-25022
ISBN 0-275-56440-1

PRAEGER PUBLISHERS
PRAEGER SPECIAL STUDIES
383 Madison Avenue, New York, N.Y. 10017, U.S.A.

Published in the United States of America in 1977
by Praeger Publishers,
A Division of Holt, Rinehart and Winston, CBS, Inc.

89 038 98765432

Printed in the United States of America

PREFACE

During 1975 and 1976, the nuclear power industry was beset by crippling uncertainties that severely limited its growth. Events in the first few months of 1977 tended to increase rather than diminish the problems afflicting the industry, and there seems to be little indication of an early reversal of a trend toward reducing the pace of nuclear expansion.

A major uncertainty plaguing the industry relates to public acceptance of nuclear power and popular concerns about the hazards of nuclear power plants. These concerns have brought a new temper to the debate over nuclear power by infusing discussion of technical and economic issues with questions of corporate social responsibility. It was through its interest in questions of corporate social responsibility and public policy that the Investor Responsibility Research Center (IRRC) first became involved in the study of nuclear power.

IRRC was founded in 1972 by a number of universities and foundations seeking impartial reporting on a variety of public policy and corporate social responsibility issues, particularly those issues being raised in shareholder resolutions at corporate annual meetings. The center now provides research reports to more than 100 institutional investors, including a number of banks, insurance companies, and other commercial institutions, as well as to universities and foundations, on a variety of controversial issues relating to the role of business in society. In the last four years, the center has prepared analytical reports on social issues ranging from coal strip mining and equal employment opportunity to political campaign contributions and overseas payments.

In 1974, at the request of its subscribers, IRRC undertook a major study of the issues involved in the debate over nuclear power. The study included a survey of consultant reports, government documents, and hearings; interviews with a wide spectrum of proponents and opponents of nuclear power; and analysis of responses to a detailed questionnaire that IRRC sent to 82 companies involved in all phases of the nuclear industry. IRRC published its initial report and updated it extensively in 1976 to provide the basis for this book.

The Center's report was designed to provide readers with a basic understanding of the primary issues providing the substance of the debate over nuclear power. It was not meant to pass judgment on the costs and benefits of nuclear power, because both are uncertain and subject to change. The objective of the study was to identify the factors likely to affect the relative costs and benefits and to provide readers with a framework against which to measure the changing aspects of the debate.

ACKNOWLEDGEMENTS

The author gratefully acknowledges support from IRRC and from its executive director, Margaret Carroll, in the preparation of this book. Special thanks are due also to Elliott Weiss for his early assistance in analysis of some of the more complex issues involved in the debate; to John B. Oliver for his research on nuclear proliferation and safeguards; to Kevin Gudridge for his work on the safety chapter; to Fred Ryland for his contribution to the chapter on regulation of the industry; to Eunice Oberg for her extensive work on many drafts of the manuscript; and to Shirley Carpenter for her supporting assistance.

On behalf of IRRC, the author would also like to thank The Ford Foundation for providing funds that have permitted publication of a paperback edition of the book.

CONTENTS

Chapter | Page

LIST OF FIGURES AND TABLES

The Nuclear Power Debate

CHAPTER

1

INTRODUCTION AND GENERAL SUMMARY

As countries around the world grope for solutions to the energy crisis, development of nuclear power to generate electricity is often touted as the best way to reduce dependence on foreign oil, as well as to ensure ultimately a limitless and economic supply of energy for the future. A growing number of countries have built nuclear plants or are actively seeking to buy them. Nations that have already developed a nuclear industry are competing increasingly to meet the demand of an expanding overseas market.

Nuclear power's attractiveness is directly related to the recognition of electricity as perhaps the most convenient and versatile form of energy. Increasingly, industry and consumers are shifting away from oil and natural gas to electricity as a source of power. As demand for electricity grows, so does interest in nuclear power. Electricity now meets 27 percent of U.S. energy needs. By the year 2000, energy officials say, electricity may account for 45 to 50 percent of all energy consumed in the United States.[1] A major portion of this electricity could come from power plants using nuclear fuel.

Until recently, there has been little question about the potential benefit to be derived from the development of nuclear power. Since the successful operation of the first commercial reactor in the United States in 1959, scientists and officials of business and government have spoken optimistically of the peaceful use of the atom to create an inexhaustible source of cheap energy. Supporters of nuclear power estimated at one time that by the year 2000 the United States would have 1,000 nuclear plants, providing more than half the country's total requirements for electricity.[2]

This optimistic vision has not won universal acceptance. Concern over the pace of nuclear development is deepening. Although the concern is shared by groups in other countries such as Sweden, France, and Japan, it has centered in the United States, where critics have begun to question the numbers of nuclear plants, the speed at which they are being built, the adequacy of facilities for storing lethal nuclear wastes, and the degree to which the United States has become involved in the marketing of nuclear plants, fuels, and technology to other countries.

Proponents of nuclear power—government, industry, and, to some extent, labor—have met increasingly strong opposition from a collection of interest groups, including consumer activists, environmentalists, some prominent scientists, a smattering of utility executives, a few politicians, and Ralph Nader, who once suggested that "if people knew what the facts were and if they had to choose between nuclear reactors and candles, they would choose candles."[3]

As recently as the end of 1973, nuclear power appeared to be gaining a certain institutional momentum. Congress, the president and the Atomic Energy Commission (AEC), manufacturers of nuclear power plants, and many of the nation's investor-owned utilities were proceeding apace with the development of nuclear power. The momentum built as the nation recognized a critical need for fuels to serve as alternatives to oil, coal, and natural gas. The imposition of the oil embargo by the Arab nations in October 1973 underscored this need.

Utilities recognized the declining availability of natural gas; with the increased price of foreign oil, the choice of what to burn in their power plants was limited to coal or nuclear fuel. Although coal is abundant, it is expensive to mine and dirty to burn. Utilities were concerned about air pollution regulations that affect their ability to use coal without substantial expenditure on devices to clean emissions. Furthermore, the nation's ability to produce coal had not improved rapidly. Several decades of debate over environmental regulations tied up vast chunks of strip-minable coal in the Western plains states.

Against this background, nuclear power was declared a critical part of the Federal Energy Administration's Project Independence, the blueprint to reduce dependence on fuels from other countries. Orders for nuclear plants grew from 14 in 1970 to 35 in 1973.[4] In 1974, President Ford urged that utilities increase the number of nuclear plants from 55 in that year to 200 by 1980.

Within two years, however, the nuclear industry had suffered a series of setbacks. It had become, in the words of Robert Seamans, director of the Nuclear Regulatory Commission, "an industry beset with problems, hobbled by self-doubts, mired in uncertainty."[5] Between 1974 and 1976, of some 200 nuclear plants under construction or on order, 24 were "removed from schedule," and another 154 were delayed for periods ranging from 12 months to several years.[6] There were only five new orders in 1975 and none in 1976. Experts projecting future development lowered their estimates from 200 units in 1980 to 180 units by 1985. Even those projections were considered overly optimistic in some quarters.[7]

The decline in nuclear development was largely the result of a changed economic situation. Demand for electricity has been drastically affected by high electricity costs resulting from the rise in coal and oil prices, a growing conservation ethic, and a mild climate. At the same time, utilities were adversely affected by high interest rates and escalating construction costs.

In addition to the economic situation, the momentum within the federal government that had spurred nuclear development was slowing. In 1975 the Atomic Commission, which had served from its inception as both promoter and regulator of nuclear power, was superseded by two agencies with separate responsibilities—the Energy Research and Development Administration (ERDA) and the Nuclear Regulatory Commission. Moreover, ERDA's function was not merely to promote nuclear power development, but to promote other sources of energy as well.

Opponents of nuclear power welcomed the slowdown in nuclear development as an opportunity to question the country's nuclear energy strategy. They sought to introduce their concerns about nuclear power into a political forum for the first time. In 1975 a number of bills challenging nuclear power were introduced in state legislatures around the country. Signatures were collected for initiative votes on nuclear power. In California and six other states, voters faced nuclear power initiatives on their ballots in 1976, and although the voters soundly defeated the initiatives, opponents of nuclear power were cheered by their success in making the development of nuclear power a political issue. In administrative hearings and in the courts, citizen groups challenged utilities' plans to finance or construct nuclear plants. Economic uncertainties delayed nuclear development long enough for opponents to raise the debate over new plants to the level of a national controversy. Ralph Nader declared that he hoped to make nuclear power the subject of the greatest citizen debate since the war in Vietnam, and in 1976 it appeared that he would be successful.

THE CONTROVERSY OVER NUCLEAR POWER

The controversy about nuclear power development spans an enormous number of issues. Most statements about nuclear power cover one or two of these and skirt the rest. However, the thrust of positions of supporters and opponents of nuclear development can be summarized as follows:

Supporters argue that U.S. and world demand for energy is growing, and that nuclear power offers the best means—outside of coal—of meeting this demand. They say that development of nuclear power is critical to the reduction of U.S. dependence on foreign oil and to the preservation of scarce domestic petroleum reserves for non-power purposes. They describe nuclear plants as safe, cheap, and reliable. Moreover, they claim that the safety of nuclear power is guaranteed by industry's programs of quality control and by stringent government regulation. Unless nuclear power is developed, they assert, the United States will most likely face blackouts, a decline in industrial growth, an increase in unemployment, and a reduced standard of living.

Opponents argue that the development of nuclear power involves extreme dangers as a result of possible power-plant malfunctions or misuse of radioactive materials. There is no need to risk these hazards, they say, because U.S. and world needs can be met economically through conservation and other sources of supply. Moreover, they contend, industry and government have been more interested in promoting than in policing nuclear power and, as a consequence, the public interest is not being protected.

THE PROGNOSIS FOR NUCLEAR POWER

The pace at which nuclear power will develop is clouded with uncertainties. There is little doubt, however, that there will be far fewer plants in operation in the next quarter century than supporters of nuclear power had once predicted. At the end of 1976 in the United States, 61 nuclear reactors were operating, representing slightly more than 9 percent of the country's total generating capacity. Another 168 reactors were either planned, under construction, or on order. Proponents of nuclear power took some satisfaction from the defeats of political initiatives which would have seriously limited further nuclear plant development there. But opponents were already rallying their forces to press legislation to limit nuclear power in a number of states around the country. Both President Ford and President Carter had called for a reexamination of the U.S. policy on exporting of nuclear materials and technology. In Congress at least five oversight committees were making plans for fresh inquiries into aspects of the nuclear industry. The Nuclear Regulatory Commission was about to begin extensive hearings into the problems of and prospects for the recycling of nuclear plant fuel—something that had been considered for many years to be an essential part of nuclear development. The Energy Research and Development Administration was working on solving problems relating to the permanent storage of waste.

Outside the United States, commitments to nuclear power grew by 17 percent in 1975 over 1974. Indonesia, Turkey, and Poland ordered nuclear plants, bringing to 41 the number of countries committed to nuclear energy. In 1976, 112 nuclear reactors were operating in 18 countries; an additional 342 plants were planned, on order, or under construction.[8]

The speed at which nuclear power will continue to grow is dependent on a number of factors: the rate at which demand for energy increases, the changing economics of alternative methods of energy production, the processes by which decisions affecting nuclear power development are made, and the degree to which they satisfy public concerns about the safety of nuclear energy. This book addresses itself to these factors, all elements in the deepening debate over nuclear power, as follows:

Economic issues: At what rate will demand for energy increase, and how can that demand be met? (Chapter 2.)
How cost-competitive are the major alternative methods of producing electricity that now exist—nuclear power and coal? (Chapter 3.)
Decision-making issues: Are the processes by which decisions to proceed with development of nuclear power, both in government and in industry, adequate to protect the interests of the public and of investors? (Chapters 4 and 5.)
Safety issues: Are nuclear power plants themselves safe? (Chapters 6 and 7.)
Can adequate safeguards be established to ensure protection against misuse of the products or by-products of those plants and to ensure the permanent safe storage of radioactive wastes? (Chapter 8.)

CAVEATS

Three caveats to readers are in order. First, many of the elements in the debate about nuclear power involve highly technical questions, such as what constitutes an acceptable level of exposure to radiological emissions, on which scientists of apparently equal eminence take opposing positions. In these situations, the researcher and the reader alike are faced with the question of which expert to believe.

A related difficulty arises even when agreement is reached on technical questions. Then, the issue often becomes what is the likelihood an event will occur, and what level of risk is acceptable: How safe is safe enough? To such questions, only subjective answers are possible.

Third, experience with civilian nuclear power plants is too limited to allow definitive conclusions. On many issues the technology is too new, the methods of analysis are too recently developed, and the operating experience of plants is too short to allow the researcher to feel sure he is in or approaching an area of reasonable certainty.

Despite these difficulties, it is possible for people other than physicists or nuclear engineers to obtain a reasonable understanding of many of the issues surrounding nuclear energy. In fact, it is because of these difficulties that many people claim a decision on nuclear power should be reached only after there is time for debate in a political forum that would allow a balancing of technical judgments against less-technical social concerns.

NOTES

1. Roger W. A. Legassie, Hearings before the House Committee on Interior and Insular Affairs, April 28, 1975, p. 118.

2. "Nuclear Power Growth 1973–2000," Wash 1139(72) (Washington, D.C.: U.S. Government Printing Office, December 2, 1971), p. 1.

3. Ralph Nader, testimony before the Commonwealth of Pennsylvania, Philadelphia, August 14, 1973.

4. "The Nuclear Industry 1974," Wash 1174-74 (Washington, D. C.: U.S. Government Printing Office, 1974), p. 7.

5. Robert Seamans, speech before the Atomic Industrial Forum annual conference, November 17, 1975.

6. Edison Electric Institute, telephone conversation, November 1976.

7. ERDA 76-1, *National Plan for Energy Research, Development and Demonstration: Creating Energy Choices for the Future,* Vol. 1 (Washington, D.C.: U. S. Government Printing Office, 1976), p. 47.

8. Atomic Industrial Forum, INFO News Release, (Washington, D.C.).

CHAPTER

2

THE ENERGY ECONOMY

The importance of nuclear power to the energy crisis turns on two related questions: To what extent will the demand for electricity grow? and How much of any growth in demand should be met by development of nuclear power, how much of it could be stemmed by conservation, and how much of it could be met by development of alternative energy sources?

Both proponents and opponents of nuclear power see at least some growth in demand. Where they differ is on how great the growth will be, on whether some of it could be stemmed without a severe impact on the economy, and on how whatever growth there is can best be met. These questions are particularly relevant to assessing the conflicting claims of industry—that continued development of nuclear power is essential—and of many critics—that a moratorium on construction of new nuclear power plants is feasible.

Predictions of total growth are important to discussion of nuclear power because it generally is acknowledged—even by opponents of nuclear power—that significant constraints exist on the amount of energy that can be supplied by fossil fuels (coal and oil). Thus, if demand for energy increases beyond certain limits, some of the increase will have to be met by other means—primarily nuclear power. The smaller the increase in demand, the greater the flexibility that will exist as to how much energy supply should come from nuclear power.

HOW MUCH DEMAND?

Just how much electricity the United States will need in the next few decades is a question of great controversy. In the past, steady growth in the demand for electricity has resulted from increasing population, declining electricity prices as a consequence of new and more effective technologies, and electricity's adaptability to a variety of uses. For more than a decade, until the oil embargo at the end of 1973, growth in demand had increased at a steady rate of around 7 percent annually. Future growth was easy to predict merely by extrapolation from past rates, and according to one utility finance officer, "All we needed was a good ruler."

Since the oil embargo, however, the electrical industry's growth projections "have been shot to hell," in the words of Paul Turner, vice president of the Atomic Industrial Forum (AIF), the nuclear industry's trade group..[1] Rising oil prices, construction costs, and interest rates forced utilities to request sharp rate increases. The rate increases and the oil embargo encouraged what has been termed "a conservation ethic" which, with the onset of a recession, cut heavily into electricity sales. Instead of the 7 percent growth rate anticipated by utilities, electricity sales in 1974 dropped slightly below the level for 1973 and in 1975 climbed only 1.8 percent above 1973 levels. In 1976, however, they rebounded to 6.3 percent above sales in 1975.[2]

These sudden shifts in consumer demand have introduced new problems into the already complex field of energy forecasting. Forecasters disagreed over whether the 1974 and 1975 experience should be interpreted as a temporary aberration or the beginning of a shift in long-term consumption patterns. They disagreed on the meaning of the increase in 1976, and they disagreed as well over the methods to be used in forecasting energy growth.

"Forecasting electricity demand and capacity requirements is an unresolved methodological problem," Duane Chapman of Cornell University has commented.[3] Past forecasts have tended to be weighted, "either consciously or unconsciously, on the basis of the prevailing conditions at the time of the forecast," according to C. A. Falcone, head of a forecast working group for the Institute of Electrical and Electronics Engineers. Forecasts in the 1950s predicting consumption in the 1970s have tended to underestimate demand; those made in the 1960s tend to run well above the levels actually reached.[4]

The proper approach to forecasting remains unclear. The Battelle Institute's Henry Hamilton notes that forecasting today requires utilities to take "new courses in uncharted waters since the techniques being recommended have not been proved . . . for load forecasting in the current environment."[5]

One major element in the debate over forecasting involves the degree to which demand is affected by price. A Federal Power Commission (FPC) task force on finance reported in 1974 that it was "unable to make a specific determination of the possible effects of price elasticity on future demand for electricity, because existing studies . . . do not provide data relevant for the future."[6] An FPC task force on conservation commented more recently that "current projections of future demand which omit price effects may seriously exaggerate the need for generating capacity."[7] Studies made over the last 15 years show a strong relationship between price and demand over the long run

—with long-run elasticity ranges between 0.6 and 1.3—but some analysts say that the relationship is far less important over the short run—0.1 over the next ten years.[8]

There is disagreement also over whether conservation of energy can reduce demand to a meaningful degree, and over what the impact of conservation would be on standards of living. Several studies support the view that conservation can play an important role in reducing energy consumption without undue adverse impact.

The Federal Energy Administration's (FEA) *National Energy Outlook*, published in 1976, states that an active energy program could reduce energy demand by "the equivalent of 3 million barrels per day in 1985."[9] FEA director Frank Zarb testified before a congressional subcommittee on energy in 1976 that "conservation is vital to our efforts to sustain our high standard of living and rekindle economic growth. Moreover, several recent analyses have shown that reducing the inefficient use of energy would not result in an employment penalty and may, in fact, create more jobs. Saving energy is synonymous with saving dollars and can, in fact, be considered as one of the least expensive energy supplies this nation has."[10] The Energy Research and Development Administration predicts that conservation of electricity alone will save 1 to 1.5 million barrels per day by 1985.[11]

The Ford Foundation's Energy Policy Project said in 1974, in its final press release, that the "nation should trim energy growth from the 4 to 5 percent of the last eight years (7 percent for electrical growth) to about 2 percent a year, and can do so without adversely affecting the economy or the amenities of our way of life. Neither jobs, nor growth rate in incomes, nor household comforts will suffer if the nation's energy growth rate is slowed by more efficient use of energy. And a decade from now, with further efficiencies and with shifts in the pattern of economic growth to less energy-intensive activities, energy growth can level off to zero."[12]

Lee Schipper, author of a report produced by the University of California's Energy Resources Group, testified in December 1975 that "we can substitute economically for at least one-third of today's energy use, with little change in lifestyle. . . ." By improving the efficiency of air conditioners, California could save enough electricity during peak hours to eliminate the need for a 1,000-megawatt power plant, Schipper said.[13] Five to eight members of a committee overseeing an FEA study of California's energy needs concluded that if the antinuclear initiatives were to pass in California, the "need for additional oil or coal for the generation of electricity in the next 20 years (if nuclear power is curtailed) could be substantially reduced or eliminated entirely if further vigorous programs to conserve energy were implemented."[14]

Industry and administration officials agree that conservation can have an important impact on consumption, but they stress the limits of that impact. FEA's Zarb argues that "even with all-out conservation beyond measures which I think the nation is today prepared to take, [the United States] can reduce that growth rate to something like 2.2 percent per year" compared with the current 3.5-percent rate. The goal of 2.2 percent, Zarb states, "is going to strain the body politic," and it is "as much as we can achieve in the next 10 years."[15]

Industry representatives say that a reduction in total demand for energy will not be matched by a corresponding drop in the demand for electricity. "In the next decade or two," said John Simpson, chairman of the Atomic Industrial Forum, in testimony late in 1975 to the California State Assembly, "even an aggressive program of conserving electric power would probably not offset the additional demand caused by existing energy uses shifting from oil and gas, let alone keep up with new demand caused by new households and rising standards of living." Energy conservation can occur, Simpson stated, "at the expense of productivity and jobs. . . . A restriction on electric power facilities would compound the serious unemployment problem . . . for all workers across the board. . . . You simply cannot have a high standard of living without a large energy supply."[16]

NUCLEAR POWER TO MEET RISING ELECTRICITY DEMAND

Energy forecasts demonstrate the interplay between the total demand for electricity and the role of nuclear power. Figure 1 shows estimates of demand for energy in 1985 and 2000 prepared in 1974 by the Energy Policy Project of the Ford Foundation and by the Atomic Energy Commission.

Nuclear power, under these projections, would increase from 0.9 quadrillion BTUs in 1973 to between 5.0 and 15.7 quadrillion BTUs in 1985. In the year 2000, as much as 78.5 quadrillion BTUs, or as little as 3.0 quadrillion, would be produced by nuclear power. If these estimates were converted to numbers of 1,000-megawatt nuclear plants, the range would be from 53 nuclear plants operating in the year 2000 under the Energy Policy Project's lowest growth projections to 1,400 plants operating under the AEC's highest. (It should be pointed out that the project's low-growth projections assume a near moratorium on new starts of nuclear power plants, and they indicate that coal production could be increased sufficiently to make up for an even slower rate of expansion of nuclear power.)

The most significant correlations in these projections are between growth in total demand and growth in nuclear power. If conservation measures held growth in demand to something approaching the Energy Policy Project's lower-level projections, it seems clear that most projected increases in development of nuclear power could be foregone. It is not surprising, therefore, that opponents of nuclear power stress the need for conservation of energy, and

FIGURE 1

Growth in Electricity Use, 1985 and 2000, as Projected by Atomic Energy Commission (AEC) and Ford Foundation Energy Policy Project (EPP) (in quadrillion BTUs)

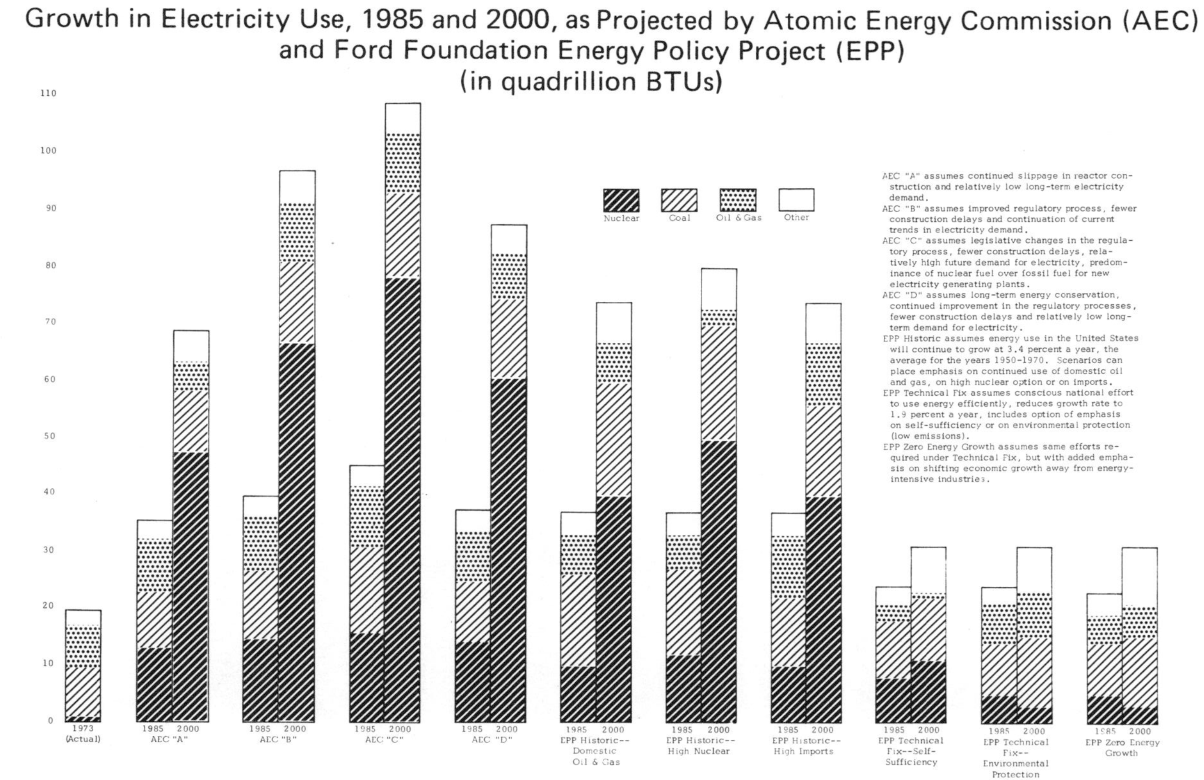

supporters of nuclear power argue that it is unrealistic to suppose that conservation measures can reduce demand sufficiently to avoid both continued nuclear power development and severe damage to the economy.

VIEWS OF NUCLEAR POWER PROPONENTS

Supporters of expanded reliance on nuclear power make two general assertions: that growth in energy supply is essential to an improving standard of living, and that nonnuclear fuel sources are either in increasingly short supply or for various reasons unacceptable to meet the demand for energy in the future. They predict a continuation in the shift to electricity from other forms of energy, and they see nuclear power as providing a virtually unlimited amount of energy—even as much as is assumed by higher-level projections—if regulatory problems can be resolved and constraints on needed material and trained manpower can be overcome.

Supply and Demand

Proponents assume, in varying degrees, steadily rising demand for energy. The demand for electricity will grow even faster, they say, because of its flexibility and attractiveness as a form of energy. Development of nuclear power, they believe, is critical to assuring an adequate supply of energy to meet that demand. And growth in energy supply, most of them assert, is an essential element in improving standards of living. Donald C. Burnham, chairman of Westinghouse, argues: "There is a wealth of data which substantiate the widely accepted contention that growth in energy usage and growth in the economy are inextricably linked."[17] Similarly, J. Harris Ward, a director of Commonwealth Edison, has stated that "the rising standard of living in the world and in the United States is related very directly to the substitution of other forms of energy for human sweat."[18] According to Burnham, a reduction in demand of the amount recommended by the Ford Foundation's Energy Policy Project would "result in substantial social upheaval, as well as economic stagnation."[19]

Alternative Fuels

Supporters of nuclear power see no way to meet required demand without an expansion of nuclear power. "In the next few decades," Burnham contends, "we truly have no practical option than to shift from our present oil-gas energy economy to an electric energy economy based primarily on coal and nuclear."[20] They cite the decline in domestic oil and gas reserves, the political and financial costs of imports, and the need to save fossil fuel for uses for which "we have no substitute: ... aircraft fuel, ... petrochemical products, ... plastics, ... lubricants."[21] And they consider it imprudent to base energy planning on the assumption that new or untested technologies, such as solar energy or nuclear fusion, can be developed on a timely basis. John Simpson of the Atomic Industrial Forum describes "fusion and solar power, especially in terms of central station generating plants ... [as] expensive pipedreams."[22]

Oil

The United States has increased its oil imports in the last few years, and a growing portion of its imports has come from countries belonging to the Organization of Petroleum Exporting Countries (OPEC). In May 1976 the United States was importing 40 percent of its total oil requirements; 60 percent of this imported oil came from OPEC countries, up from 49 percent at the time of their oil embargo in 1973–74. "By 11 percent," Zarb of FEA says, "we have become more vulnerable to another disruption from that part of the world."[23]

Proponents of nuclear power are concerned about this growing reliance on foreign oil and point out that by building 200 nuclear plants the United States could save 6.25 million barrels a day—almost equal what it now imports.[24] And, they say, a shift to nuclear power would benefit the U.S. balance of payments. The United States will spend an estimated $30 billion on oil imports on 1976.[25]

Coal

As between nuclear power and coal, nuclear power is seen by its supporters as clearly the safer, the less damaging to the environment, and the more plentiful in the long run. In June 1974 a group of six scientists and engineers —several of whom previously had worked with groups opposing nuclear power—issued a paper called "The Nuclear Debate: A Call to Reason." They stated that electricity systems based on surface- and deep-mined coal pose the greatest risk in terms of occupational safety, large-scale accidents (such as slag slides), and environmental impact, and nuclear and oil-based systems pose the smallest. "Rigorous examination of the present risks, costs, and impact of all electrical power sources," they said, "lead us to conclude that nuclear power is more than acceptable; it is preferable."[26]

Industry officials cite the dangers from mining coal—more than 100,000 miners have died since 1900, according to energy consultant Ralph Lapp[27]— and from burning coal. Bernard Cohen, professor of physics at the University of Pittsburgh, states that "a typical coal-fired power plant probably kills about

50 people per year with its air pollution, whereas a typical nuclear power plant of the same capacity probably kills an average of 0.01 people per year."[28]

There is some support outside of industry for claims concerning the hazards of coal. A recent study by C. L. Comar and L. A. Sagan, of the utility-supported Electric Power Research Institute and the Palo Alto Medical Clinic, respectively, concluded that "occupational deaths from the use of coal are considerably greater than for the three other classes of fuel"—natural gas, oil, or nuclear power. Premature deaths among the general public are very much more likely to result from the use of coal and oil than from natural gas or nuclear fuel"[29] as a result of their emissions. And concerns with pollution from coal plants have led some environmentalists to support nuclear power. Lapp was formerly head of the Sierra Club's energy policy committee. Larry I. Moss, former president of the Sierra Club, says that he supports nuclear power because he believes that the alternative is to kill many more people with air pollution from fossil fuel plants. Moss believes it will be easier to develop a system to minimize the hazards of nuclear power wastes than it will be to control air pollution from fossil fuel plants.[30]

Utility managers and others also question whether sufficient amounts of coal can be made available on an assured basis. They point out that in 1973 the United States produced 592 million tons of coal, 12 million short of what the utilities needed. They hoped for a 5 to 6 percent increase in 1974, but a strike by West Virginia miners stopped production for 28 days and dropped total production for the year at least 30 million tons below demand. Production in 1975 improved only slightly. For coal to replace uranium by the year 2000, according to some observers, would require more than 2 billion tons annually, the equivalent of 100,000 railroad carloads every day. Many people consider such growth unlikely.

"For all practical purposes," says Kenneth Davis of Bechtel, "the mining of coal has been essentially constant for the last 60 or more years. Coal still faces significant constraints on growth . . . There are uncertainties in regulations governing surface mining, questions on environmental regulations, and somewhat more uncertainty in labor stability and mine safety. There are additional problems in the methods of transportation of increased volumes of coal to various parts of the country. There are serious questions about the burning of coal in many areas."[31]

VIEWS OF NUCLEAR POWER OPPONENTS

Nuclear opponents range from those who urge a slowdown of nuclear power development to those who urge an outright moratorium. In general, opponents assert that top priority should go to conservation of energy and to development of nonnuclear technologies.

Supply and Demand

Most critics contend that conservation measures can be implemented to reduce demand for energy to levels that can be met without additional development of nuclear power. And, they say, the reduction in demand can be made without seriously affecting the economic well-being in the country. In support of this conclusion, they cite studies such as those by the Ford Foundations's Energy Policy Project by the University of California's Energy Resources Group mentioned in the first part of this chapter.

Overdependence on Nuclear Power

Critics assert that nuclear power is not fully enough developed to provide a sure substitute for fossil fuels, that it cannot be developed at the rates its proponents project, and that other fuels can be used to fill the gap between energy available and energy needed.

Senator Abraham Ribicoff (D-Conn.) wrote in 1974 in the New York *Times:*

> We are now nearly 30 years into the nuclear age, but still not near the point where we can safely harness the peaceful atom as a major substitute for oil.
>
> By placing most of our energy research-and-development eggs in the nuclear basket, we now find ourselves with problem-prone nuclear power as the only alternative to oil and coal as our major source of energy.
>
> Instead of giving us a genuine energy option, the AEC and the nuclear industry have sealed our dependence on blackmailing Arab sheiks and price-gouging multinational corporations for oil. . . .
>
> The time has come to follow a bold energy research-and-development policy that is faithful only to the independence of invention. Continuing to serve basically entrenched nuclear power interests could leave us dangerously dependent on oil and other fossil fuels for our energy.[32]

Daniel Ford, executive director of the Union of Concerned Scientists, agrees with Ribicoff's concerns. In testimony before the House Interior Committee in 1975, Ford argued that

> if the administration's proposal for nuclear energy is fulfilled, by 1985 we will have developed major dependence on nuclear power amounting to around 25 percent of all of our electricity projected. At that point, shutting off or restricting nuclear power plants in operation for one of these emergency safety inspections could have very serious repercussions. Those of us who learned the lesson from the Arab oil embargo in 1973 of the economic difficulties of relying on a single unreliable source should also see that the

special safety precautions that have to be applied to the nuclear program may make it unreliable from the point of view of the over-all economy.[33]

Others argue that projections of the rate at which nuclear power will expand have been and continue to be unduly optimistic. Barry Commoner argues that new information on nuclear power's "technical feasibility, reliabiltiy, environmental impact, safety and economic soundness . . . shows that in all these aspects the nuclear power system falls so far short of the original optimistic assumptions that there is very little likelihood that it can continue beyond the next decade or so."[34]

Alternative Fuels

Critics view the energy problem primarily as one of filling, for a period of 10 or 20 years, the gap created by reductions in supplies of cheap oil and gas. Then, they say, reliance on yet-to-be-disclosed technologies, such as solar energy or nuclear fusion, will become feasible. Commoner states, "It is now clear that even on a national scale, and certainly in California, solar energy sources could totally replace nuclear power. It is also clear that the production of solar collectors would involve much greater employment opportunities and for a wider range of trades than nuclear power plants."[35]

Most critics assert that coal can fill any energy needs that remain unmet by other fuels. The environmental and human costs of a coal-based energy cycle, they argue, can be minimized by directing increasing attention to mine safety, requiring reclamation of strip-mined lands and utilizing a variety of technologies to reduce air pollution caused by burning coal.

NOTES

1. Interview with Paul Turner, June 2, 1974.

2. Edison Electric Institute figures.

3. Duane Chapman, National Power Survey Task Force on Conservation and Fuel Supply of the Technical Advisory Committee on Conservation of Energy, *Power Generation: Conservation, Health and Fuel Supply* (Washington, D.C.: U.S. Government Printing Office, March 1975), p. 9.

4. C. A. Falcone, *Electric Energy Technology Forecast,* a paper prepared by the Energy Forecasting Group of the Institute of Electrical and Electronics Engineers, January 1975, p. 2.

5. Henry R. Hamilton, "Forecasting Electricity and Energy Demand" (Batelle-Columbus, Columbus, Ohio), n.d., p. 4.

6. Technical Advisory Committee on Finance, Federal Power Commission National Power Survey, *The Financial Outlook for the Electric Power Industry* (Washington, D.C.: U.S. Government Printing Office, December 1974), p. 131.

7. Technical Advisory Committee on Conservation of Energy, op. cit., p. 31.

8. Frost and Sullivan, Inc., *Fossil Fuel Electric Generating Station Requirements 1976–1985,* (New York, 1976), p. 15 ff.
9. Federal Energy Administration, *1976 Executive Summary, National Energy Outlook* (Washington, D.C.: U.S. Government Printing Office, 1976), p. 6.
10. Frank Zarb, testimony before the Senate Subcommittee on Energy, Joint Economic Committee, February 3, 1976, p. 3.
11. Energy Research and Development Administration, Office of Public Information, December 1976.
12. Ford Foundation, *A Time to Choose,* press release, November 1974.
13. Lee Schipper, *Hearings on Proposition 15,* Vol. XIV, "General Hearings Part 1," California State Assembly, December 9, 1975, pp. 26, 27.
14. Joint Report on Five of the Eight Members of the Oversight Committee, *Direct and Indirect Economic, Social, and Environmental Impacts of the Passage of the California Nuclear Power Plants Initiative,* Executive Summary, April 1976, reprinted in *Hearings on Proposition 15,* Vol. XVI, California State Assembly, May 14, 1976, p. 127.
15. Frank Zarb, *Hearings on Proposition 15,* Vol. XVI, op. cit., p. 30.
16. John Simpson, *Hearings on Proposition 15,* Vol. XIV, op. cit., pp. 4, 5.
17. Donald C. Burnham, in *A Time to Choose* op. cit., p. 367.
18. J. Harris Ward, in ibid., p. 409.
19. Burnham, op. cit., p. 367.
20. Ibid., p. 366.
21. John W. Simpson, "A World Without Natural Fuels?" Westinghouse advertisement in *Fortune,* June 1974, p. 32 ff.
22. Simpson, *Hearings on Proposition 15,* Vol. XIV, op. cit., p. 3.
23. Zarb, *Hearings on Proposition 15,* Vol. XIV., op. cit., p. 29.
24. George Stathakis, "There is no reason in the world why nuclear critics should be more convincing in talking about our industry than we are," GE's *Nuclear Power Newsletter,* Winter 1974, p. 1.
25. FEA estimates, November 1976.
26. Ian Forbes, et al, "A Call to Reason," published by California Council for Environmental and Economic Balance, September 1975, p. 1.
27. Ralph Lapp, "Nuclear Salvation or Nuclear Folly," *The New York Times Magazine,* February 10, 1974, p. 73.
28. Bernard L. Cohen, "Perspectives on the Nuclear Debate," *Bulletin of the Atomic Scientists,* October 1974. p. 35.
29. C. L. Comar and L. A. Sagan, "Health Effects of Energy Production and Conversion," *Annual Review of Energy,* Vol. 1, 1976, pp. 197, 598.
30. Larry I. Moss, interview with author, May 1974.
31. W. Kenneth Davis, *Hearings on Proposition 15,* Vol. II, December 10, 1975, p. 79.
32. Abraham Ribicoff, New York *Times,* December 5, 1974.
33. Daniel Ford, testimony before the Subcommittee on Energy and the Environment, House Committee on Interior and Insular Affairs (Washington, D.C.: U.S. Government Printing Office, April 28, 1975), p. 189.
34. Barry Commoner, *Hearings on Prosposition 15,* Vol. XV, October 15, 1975, pp. 123, 124.
35. Ibid., p. 131.

CHAPTER

3

THE ECONOMICS OF NUCLEAR POWER

The pace and extent of nuclear power plant development depend largely on the degree to which nuclear-generated power appears to be economically competitive with energy from coal-fired power plants. There is considerable debate over the relative economics of coal and nuclear power, and predicting the costs of future energy generation is fraught with uncertainties that are both technical and political. Factors such as the capacity at which plants are able to operate or the cost and efficiency of equipment required to meet safety and environmental standards will play a particularly important role in determining the relative competitiveness of coal and nuclear power plants.

Supporters of nuclear power argue that it provides considerable savings to consumers. The Atomic Industrial Forum (AIF) says its survey of utilities during the first half of 1976 demonstrated that nuclear power is the lowest-cost producer of electricity. Nuclear plant operation during this period, the AIF maintains, accounted for "estimated fuel savings of over 5 billion gallons of oil or 27 million tons of coal and [produced an] estimated generating-cost savings of $625 million."[1] Customers at Northeast Utilities saved $39.1 million by having electricity generated by nuclear rather than oil-burning power plants, according to the forum's figures.

Many proponents say that nuclear power will hold an advantage over alternative fuel sources in the future. The utility-funded Edison Electric Institute (EEI) recently stated that "nuclear power through 1990, at least, is expected to be more economical than coal-fired generation in many areas of the country."[2] The EEI press release declares that figures for a recent survey of utilities show "an over-all advantage in favor of nuclear power of about 30 percent."

Increasingly, critics are challenging these claims. "Nuclear plants have more problems than a hound dog has fleas," Donald Cook, who was chairman of American Electric Power Company until 1976, has commented. "There are problems associated with bringing a nuclear plant into existence, in getting enrichment of the fuel and reprocessing of spent fuel. This, coupled with the length of time and investment represented only by work in progress and no income, presents a set of difficulties as to lead any prudent person to go for a conventional coal-fired plant."[3] American Electric Power plans to rely almost entirely on coal-fired power plants.

The critics argue that the claims made by nuclear power enthusiasts are exaggerated or misleading. In August 1975 the Council on Economic Priorities (CEP) issued a report disputing a statement by Consolidated Edison that its nuclear plants had saved customers $49 million in 1974. CEP maintained that Con Ed's computations failed to take into consideration the high capital cost for constructing nuclear plants. The capital cost and problems with maintenance offset any gains from the lower cost of nuclear fuel, CEP concluded, and had the company built a coal plant instead, the company could have saved its customers $26.8 million.[4] (Con Ed spokesmen responded later, saying their figures related only to saving on fuel costs.)

Even if nuclear power is competitive, nuclear opponents say, it has become so only because large federal grants have been made available. Ron Lanoue of Ralph Nader's Center for Study of Responsive Law has described the nuclear industry as being "on the welfare rolls. . . . Most economic estimates of the cost of nuclear power are biased because they do not include the cost of the subsidy programs that our tax dollars underwrite."[5] End these subsidies, critics say, and then we will be able to see if nuclear power is competitive.

The total cost of electricity as it leaves the power station, referred to as the busbar cost, is the product of several factors: (1) the construction cost of the plant, including hardware, labor, the original capital borrowed, and the interest generated on that capital, and inflation on capital costs: (2) the cost of operating and maintaining the plant; (3) the cost—usually reflected under capital, operation, and maintenance or fuel costs—of meeting environmental and safety standards; and (4) the cost of the fuel itself. The Energy Research and Development Administration estimates the per-kilowatt cost of electricity generated from 1,000-megawatt nuclear, coal and oil plants beginning operation in 1980 as follows:[6]

As the table shows, the capital costs constitute a large portion of the total costs of generating electricity at a nuclear plant than they do at a coal-fired plants. Construction of a nuclear plant may cost several hundred million dollars more than construction of a fossil fuel plant. Nuclear fuel, on the other hand, is far less costly than coal or oil, and savings from fuel may enable a nuclear plant to compensate for more than its initial high capital costs, providing that the plant is able to operate efficiently over an extended period of time.

	Electricity Cost (in Mills per Kilowatt Hour)		
	Nuclear Plant	Coal Plant	Oil
Capital costs	18.7	15.2	10.5
Fuel Costs	5.8	13.7	25.7
Operation and maintenance costs	2.8	3.3	2.2
Total costs	27.3	32.2	38.4

In the final analysis, then, the critical determinant of busbar costs is the efficiency at which a plant is able to operate. Estimates of generating costs assume operation of a plant for a certain number of years—usually 30—at an estimated capacity level—60 to 80 percent. These assumptions, however, are being drawn from an extremely limited data base. The history of operating nuclear and large-scale coal-fired plants is short. For example, although industry spokesmen frequently mention 1,500 reactor-years of operating history with nuclear power, that total includes operation of small military reactors, reactor-powered submarines and aircraft carriers, and civilian reactors in power plants far smaller than those being constructed today or planned for the future. Of 61 nuclear plants operating today, 33 are 700 megawatts or larger in size, and their combined operating history covers only 76 reactor-years. This small sample can be influenced strongly by the good or bad performance of just one plant. Thus, the available data may present too optimistic or too pessimistic a preview of what is likely to occur in the future.

The following sections of this chapter outline the various component costs of power and compare the factors affecting these costs at coal and nuclear power plants. Discussion of costs is limited to coal and nuclear power because the cost of generating electricity by burning oil is so much higher than these two sources and because the U.S. government has begun policies to discourage construction of new oil and gas-fired plants.

CAPITAL COSTS

One official of the Atomic Industrial Forum has noted that "estimating capital costs for power plants is like shooting at a moving target." The difficulty of the task is illustrated by Figure 2 which compares estimates made by the Atomic Energy Commission (AEC) from 1967 to 1975 of the cost of building 1,000-megawatt nuclear and coal-fired plants. With respect to both types of power plants, the costs estimated in 1974 were more than five times those estimated in 1967.

This history of cost increases, and the large proportion of total estimated costs that are represented by assumptions about cost escalation and interest costs, contribute to uncertainty about the accuracy of any predictions about future costs. The uncertainty is borne out by the experience of some individual

FIGURE 2

Comparison of Cost Estimates for Nuclear and Coal-Fired Plants
(total investment cost for 1,000-MWe units)

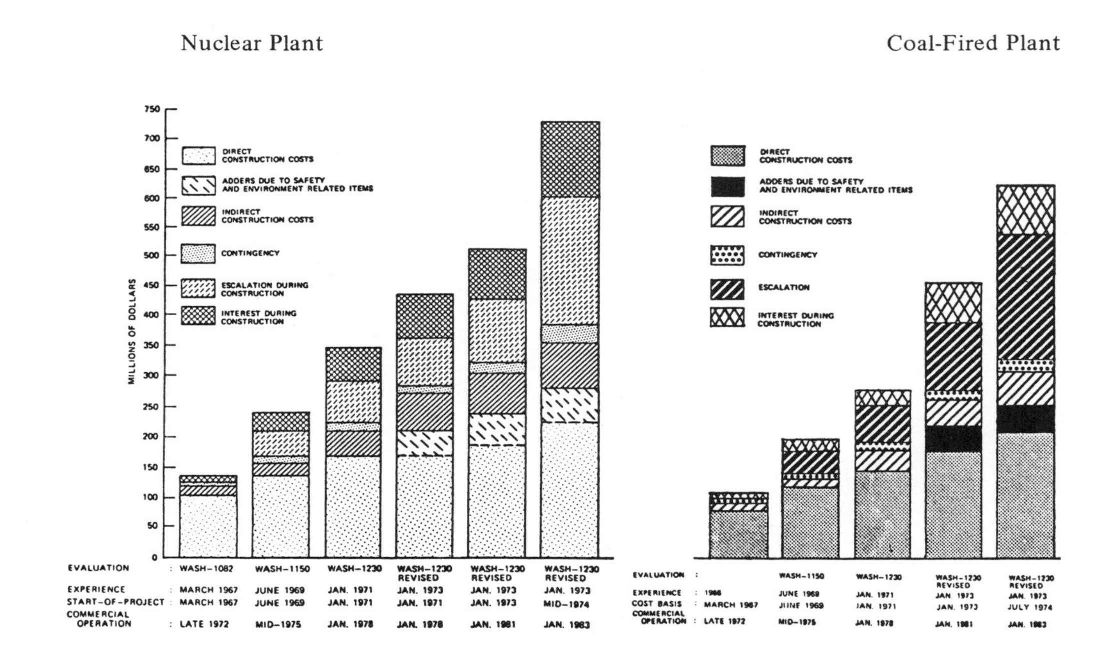

Source: Atomic Energy Commission reports.

utilities in building power plants in the past. William Rosenberg of the Michigan Public Service Commission has testified before the Senate Interior and Insular Affairs Committee that "we have seen . . . a lot of cost-plus planning and cost-plus rate-making. The cost overruns on construction projects nationwide run into billions of dollars. . . . One particular (nuclear) plant . . . came in 287 percent over budget."[7] Charles Pierce, president of the Long Island Lighting Company, noted that the Shoreham nuclear plant, which was expected in 1969 to cost $278 per kilowatt of capacity to build, was expected in 1974 to cost $848 per kilowatt.[8] And in December 1975, Robert Kirby, president of Guardian Capital Trust Company, told a meeting of the Atomic Industrial Forum that "the costs of constructing such a [nuclear] facility have gone off the chart and still appear to be rising so rapidly that no one in his right mind could guess what the actual costs of a new plant might be when that plant was completed."[9]

According to Irwin Bupp, associate professor at the Massachusetts Institute of Technology, "on average [nuclear] plants in the United States which are complete or nearly complete have cost on the order of twice as much in real (constant) dollars, to build as at the time they were ordered."[10] Costs will continue to escalate in the future. General Electric's (GE) office of strategic planning estimates that the costs of nuclear plants will more than double between 1975 and 1984.[11] These are GE's figures:

GE's figures assume:

GE's Estimates of Costs of Construction, 1975 and 1984

	Nuclear Power Plants		Coal-Fired Plants	
	Millions of 1975 Dollars per Kilowatt		Millions of 1975 Dollars per Kilowatt	
	1975	1984	1975	1984
Hardware and construction costs	230	325	265	315
Professional services and other indirect costs	85	118	60	90
Interest during construction	70	230	55	210
Escalation of costs during construction	80	280	60	235
Total	465	953	440	850

The nuclear plant will be constructed in nine years, the coal plant in 6.5 years.
The interest during construction will be 9 percent.
The escalation will be 7.5 percent to 1980, 5 percent thereafter.

The construction costs for the coal plant include $80 per kilowatt for stack-gas scrubbing equipment to remove sulfur dioxide from plant emissions.

GE's figures demonstrate that the cost of nuclear plants is increasing slightly faster than that of coal plants. The cost for coal plants, excluding interest and escalation, will rise 25 percent between 1975 and 1984, and that for nuclear plants, excluding interest and escalation, will increase 41 percent in the same period. Any further increase in these costs, even if applied equally to coal and nuclear plants, is apt to improve the competitiveness of coal because of the longer period required for construction of a nuclear plant.

One change that could affect the competitive position of nuclear power would be a shortening of the time required to license and construct a nuclear plant. Figure 2 shows that more than half the costs of a nuclear plant in 1984 can be attributed to interest during construction and to escalation. A reduction in the time between initiating plans for the plant and putting it into operation would cut these costs. Both industry and government officials hope to be able to reduce this period from the nearly 11 years required today to 6 years by standardizing plant design. But reform of the regulatory and licensing process is difficult, and many observers are pessimistic about its chances. Meaningful reform of the process probably will not take place, said then—former Interior Secretary Thomas Kleppe, "until the lights go out in San Diego."[12]

The competitive position of coal plants, on the other hand, would be affected by technical advances that alter the cost of stack-gas scrubbing equipment. The equipment, required in many parts of the country to enable coal emissions plants to meet air quality standards, represents a major portion of the capital costs of a new coal plant. (GE assumes a cost of $80 million for stack-gas scrubbing equipment in 1984; others assume a cost of $100 million or more.) In fact, the cost of scrubbers is uncertain. The author of an article on the subject, Sam Ruggeri, describes the uncertainties this way: "Talk to any two people on scrubber costs and you get disagreement, except for a clear-cut trend showing that regulatory authorities and vendors quote lower figures than do the utilities, who actually pay the costs."[13]

OPERATION AND MAINTENANCE COSTS

Federal Power Commission figures for 1975 show that nuclear plants were slightly more expensive than coal plants to operate and maintain. Operation and maintenance costs averaged 2.7 mills per kilowatt hour for nuclear plants, 1.5 for coal plants. The difference may be attributed to the more sophisticated safety equipment required at nuclear plants and to the fact that few coal plants have installed scrubbing equipment, a step assumed to increase operation and maintenance costs at coal plants by 1.5 to 2 mills per kilowatt hour.[14]

Another factor affecting maintenance at nuclear plants is the strict limitation on radiation exposure of workers. The limits allow workers only slight exposure for a very brief period of time, after which they are considered "burned out" and must be replaced by other workers. As a result, companies at nuclear plants may have to train several workers to perform the same kinds of repairs, and repairs in areas where there is radiation exposure may take more time and workers than would be the case for similar repairs in nonnuclear plants.

FPC figures seem to indicate that estimates of operation and maintenance (O&M) costs made earlier by the AEC are too low. In 1974 the AEC estimated that O&M costs in 1982 would be 0.8 and 2.0 mills per kilowatt hour (kwh) for nuclear and coal plants, respectively. O&M costs at nuclear plants have already exceeded this estimate, and costs at coal plants are likely to do so once more stack-gas scrubbers have been installed.

Charles Komanoff of the Council on Economic Priorities estimated in testimony before the New Jersey Public Utilities Commission that annual O&M costs for a nuclear plant during the period 1983–92 would be 5.3 mills per kwh. O&M costs for coal plants during the same period could run 10.5 mills per kwh for plants with scrubbers and 5.3 mills per kwh for plants without scrubbers, Komanoff said.[15]

FUEL COSTS

It is in the area of fuel costs that both proponents and opponents of nuclear power agree that nuclear plants have a decisive advantage over coal-fired units. The FPC reported recently that the fuel costs of nuclear plants in 1975, including uranium enrichment and fuel fabrication, were 2.7 mills per kwh; fuel costs at coal-fired plants averaged 9.5 mills per kwh.[16] And Komanoff, assuming a 6 percent rate of escalation, calculates that the average fuel cost for the first ten years of operation of a nuclear plant starting up in 1983 will be 12.8 mills kwh, and the cost of burning low-sulfur coal will be close to 20 mills per kwh.[17]

It appears for the moment as though the price of uranium is rising more rapidly than the price of coal. Coal prices have been relatively stable over the last year. The cost of uranium, however, has moved rapidly from an estimated $11 per pound of uranium oxide in 1975 to more than $50 per pound for delivery in 1980. According to Sargent and Lundy, an engineering firm, "quotations for delivery beyond 1980 are virtually unobtainable."[18] ERDA estimated costs in 1976 were as follows:[19]

Component of Nuclear Fuel Cost	Cost (mills per kilowatt hour, 1976)
Uranium ore concentrates	3.7
Conversion	0.1
Enrichment of uranium	2.5
Fabrication of fuel assemblies	0.5
Transportation and reprocessing of spent fuel; waste management	1.0
Credits for recycle of uranium and plutonium recovered from spent fuel	–2.0
Total	5.8

Not only has the cost of uranium gone up considerably in the last two years, but so has the cost of enrichment. The 1976 figures demonstrate that enrichment costs have more than doubled since 1974. The total cost of nuclear fuel, assuming no credit for recycling of plutonium or uranium from spent fuel, has gone up almost 300 percent.

A number of factors could affect the future cost of fuel for coal or nuclear power plants. The costs of certain of the components of nuclear fuel remain very much undetermined. No precise figures can be assumed for the cost of reprocessing or waste management, or of the potential return credit from plutonium or uranium extracted from spent fuel (the fuel products removed from a reactor after use). Most knowledgeable observers expect that the cost of reprocessing or waste disposal will be relatively minor. But only limited commercial reprocessing has taken place, and no permanent waste storage has ever been done, and it is possible that restrictions on reprocessing to ensure safety and safeguard wastes will raise these costs beyond current estimates. Sargent and Lundy state that estimates of future reprocessing and waste disposal costs are "nearly meaningless"; although "based on the best information available to us, values as high as 10 times the 1969 levels have been used in recent estimates."[20]

In addition, the price of uranium itself may continue to rise, as it becomes more scarce and as the prices of oil and coal climb. There is some debate over just how much uranium will be available after 1980. Ray Dickeman, president of Exxon Nuclear, has stated, "It is probably too late to avoid an important uranium import program;"[21] but the availability of foreign reserves, in quantities at an acceptable price, is also clouded. The *Wall Street Journal* reported in June 1976 that "dozens of utilities are delaying or canceling nuclear power plants, not only because of the recently low growth in demand for electricity and the hassles they face over environmental protection, but also because of the uncertainties over uranium supplies."[22] And Westinghouse, unable to find adequate supplies of uranium at what it considered to be a reasonable price, backed away from contracts to supply some 27 utilities with 65 million pounds of uranium over the next 20 years. The utilities, facing uranium shortages, have sued Westinghouse to uphold the contracts.[23]

Another factor that could affect the demand for uranium is the efficiency with which reactors are able to burn uranium fuel. Reactors have been designed to produce a certain amount of power per ton of uranium burned. Some nuclear critics, most notably Jim Harding of the California State Energy Resources Conservation and Development Commission, have suggested that plants have been burning uranium fuel at a rate well below design capacity. Inefficient burning of uranium will require utilities to burn more fuel, thus increasing the demand for uranium and raising utility fuel bills.[24]

Nuclear industry and government representatives deny that utilities will face shortages in the near term, and a report issued recently by the Federal Energy Resources Council stated that "the adequacy of uranium to provide fuel (over their 30-year lifetime) for all existing, planned and additional reactors which may be placed into service by 1990 is a reasonable national planning assumption." The council, however, acknowledged that "the possibility does exist for short-term uranium supply problems if timely industrial decisions to expand domestic uranium mining and milling capacity are not forthcoming. Delays in adding enrichment capacity and delays in achieving recycle of uranium or plutonium would increase the possibility of such a shortage."[25]

Similarly, a number of factors could affect the future cost and availability of coal. Despite exhortations for increased coal production, total coal tonnage produced in the last two years has remained relatively static. Deep-mined coal production is highly susceptible to strikes; strip-mined coal production may be affected by legislation requiring strict observance of standards for reclaiming strip-mined land. The cost of coal to certain utilities will also be affected by plant location—its proximity to suitable supplies of coal—and by the strictness of the air quality standards set for the region in which the power plant is to be located. Air quality standards will determine the type of coal a utility is able to burn. They may require companies to incur heavy transport costs in order to obtain sufficient quantities of low-sulfur coal or additional costs to clean coal before burning or, using stack-gas scrubbers, to clean the plant's air emissions.

SUBSIDIES

In estimating the cost of electricity from nuclear- and coal-fired power plants, neither industry representatives nor ERDA officials mention the government funds used to support development of these two power sources. This is quite understandable, in that most estimates are designed to show costs to the utilities, but the omission does highlight an argument made by critics of nuclear power—that the competitive viability of nuclear power exists primarily because of the large supply of government funding which has been provided to the program.

Observers estimate that more than $55 billion has been spent to date on development of civilian nuclear power. Of this total, some $5 billion has been provided by the federal government. ERDA has proposed to spend $1.3 billion on civilian nuclear programs in fiscal year 1977, and the administration has approved additional programs which, if supported in Congress, could greatly increase public support to the nuclear industry.[26]

Public funds are being used today in a number of areas, including research on safety, safeguards, and storage; new reactor demonstration projects; and experiments involving recycling and the fabrication of mixed oxide fuels. ERDA continues to control nuclear fuel enrichment, and the government continues to support a program of federally subsidized insurance to the nuclear industry through the Price-Anderson Act. (See Chapter 4 for a full description.) New programs suggested by the Ford administration included federal loan guarantees of up to $8 billion to private corporations interested in building new uranium enrichment facilities and assistance in the development of a demonstration reprocessing plant that could cost more than $1 billion. Both programs have encountered criticism, and neither passed Congress in 1976.

A number of critics argue that government assistance in these programs distorts market judgments of the costs of nuclear power. They maintain that if it were not for these subsidies, or if comparable amounts were available to support other fuel cycles, nuclear power would cease to be an economically viable source of energy supply.

ERDA contends that its policy, and previously that of the Atomic Energy Commission, has been to use public funds only as long as necessary in areas where private funds are not available. Its aim, it says, has been to demonstrate that certain aspects of the nuclear industry are commercially viable. Once that is demonstrated, ERDA says, it expects industry to take over.

Industry also supports use of government funds for these programs. Spokesmen argue that government assistance is appropriate in support of national objectives of lowering air pollution for burning coal and decreasing imports of oil. The government, company representatives say, should assist industry in areas where private companies are forced to assume abnormal risks or expenditures as a result of these public policy decisions. Westinghouse argues that any cost accounting considering the subsidy to nuclear power should also consider "subsidies to the coal, oil and natural gas industry." AIF officials mention the black lung disability program for coal miners and tax credits under the oil depletion allowance as forms of government subsidy to fossil fuel producers. Southern California Edison contends that "about one-third of civilian nuclear expenditure is recoverable" through sales of enriched uranium, licensing fees, and taxes on nuclear power plants and supporting facilities.[27]

OPERATING CAPACITY

Regardless of fuel, O&M, or capital costs, it is quite likely that the chief determinant of nuclear power's competitiveness with coal will be the capacity at which the nuclear plants are able to operate over the life of the plants. By operating efficiently, a nuclear plant will be able to take advantage of its lower-cost fuel and overcome its initial heavy capital cost.

Electric utilities, in general, have three kinds of power plants, which operate to meet three levels of demand: base load, intermediate load, and peak load. Base load is the minimum level of electricity demand that exists in a system around the clock, all year long. Consequently, base-load power plants are designed to run almost continuously at full capacity to meet this level of demand. Intermediate-load plants are designed to operate frequently, supplementing the output of base-load plants to meet increases in demand, above base-load levels, that occur with some regularity. Peak-load plants operate only occasionally, when demand in a system reaches unusually high levels. For example, peak demand often occurs on very hot summer days, when many air conditioning units are operating.

Nuclear power plants—and large coal-fired plants—are used almost exclusively to meet base-load requirements. The high capital and low fuel costs of nuclear plants render use of them for other than base-loading purposes very expensive, because the savings in fuel costs are modest, and the amount of capital costs to be attributed to each kilowatt hour of output becomes very high. For much the same reasons, it is important that nuclear power plants built for base-loading purposes operate at close to full capacity if they are to produce electricity at a competitive cost.

In evaluating the past performance of nuclear power plants, it is important to distinguish between two measures of performance frequently used: availability and capacity. "Availability" describes the time that a plant is available for operation at some level. To say a plant was available does not mean that it actually operated during the time, or that it was available for operation at full capacity—only that it was available to meet some level of demand. Hence, a 1,000-megawatt plant that produces electricity during nine months of the year and is out of operation for three months would have an availability factor of 75 percent, regardless of how much electricity it actually produced. Thus, the availability factors of coal or nuclear plants are less relevant to determining their economic competitiveness than are the capacity factors. "Capacity factor" describes the level at which a plant actually has operated. A 1,000-megawatt plant producing 800 megawatts of power 100 percent of the time or 1,000 megawatts 80 percent of the time is operating with an 80 percent capacity factor. Because nuclear power plants must shut down an average of one month a year for refueling, the maximum realizable capacity factor of a nuclear plant is 92 percent.

Past Performance

Government officials, reactor vendors, and many utilities have referred, in the past, to an 80 percent capacity factor as their performance goal. Until recently, they often assumed a 75 to 80 percent capacity factor in estimating the costs of generating electricity.[28] To date, nuclear power plants have generally operated with a capacity factor considerably below 80 percent, although at about the same level as large-scale fossil fuel plants. (The operating history to date is shown in Table 1.) A report from the Edison Electric Institute showed fossil fuel plants, 6000 megawatts or above in size, averaged a 58 percent capacity factor between 1965 and 1974. Nuclear plants were rated at 59.61 percent.[29] Atomic Industrial Forum figures for the first half of 1976 showed base-loaded coal plants with a capacity factor of 59 percent compared with 58.1 percent of nuclear plants.[30] Of 53 nuclear plants surveyed by the Nuclear Regulatory Commission in 1976, 20 had a capacity factor below 56 percent.[31] The breakdown was as follows:

Capacity Factor	Number of Plants
Below 33 percent	6
33–56 percent	14
56–66 percent	19
Above 67 percent	14

Interpreting the Data

The significance of plant performance statistics is subject to some dispute, primarily because the data base is rather small. Experience with nuclear plants of the size being constructed today—1,000 megawatts or greater—is extremely limited. Supercritical coal plants—which operate at a higher temperature and with a higher steam pressure, thus increasing their efficiency—have only been in operation since 1968.

Advocates of nuclear power argue that as times goes by, the industry's record will improve. ERDA estimates that new nuclear plants experience a three-to-four-year shakedown period, after which time a higher factor can be expected. Commonwealth Edison, for example, has stated that while "the combined average for our 800-mw nuclear units from 1972 through 1974 is 60 percent, it is expected that this value will increase to about 65 percent as continued operation eliminates the starting troubles typical of any new generating unit."[32] The New Jersey Public Service Electric and Gas Company estimates that plants going into operation in 1983 will have a capacity factor for the first two years of 65 percent, rising to 74 percent in the third year, and to 78 percent in the sixth year.[33] Industry spokesmen explain nuclear power's

Table 1

U.S. Nuclear Plant Operating Experience

Plant	Utility	Date Began Operation	Megawatts	Unit Capacity Factor to August 1976[a]
Arkansas One 1	Arkansas P&L	1974	850	55.5
Beaver Valley	Consumers Light	1975	852	—
Big Rock Point	Duquesne Light[b]	1965	72	52.4
Browns Ferry 1	TVA	1974	1,098	25.5
Browns Ferry 2	TVA	1974	1,098	13.5
Browns Ferry 3	TVA[b]	1975	1,065	—
Brunswick 2	Carolina P&L	1975	821	41.9
Calvert Cliffs 1	Baltimore G & E	1974	845	85.7
Calvert Cliffs 2	Baltimore G & E[b]	1977	845	—
Conn. Yankee	Connecticut Yankee Power	1968	575	79.5
Cook 1	Indiana Michigan Power	1975	1,060	80.1
Cooper	Nebraska Public Power Dist.	1974	800	57.7
Dresden 1	Commonwealth Edison	1960	200	50.0
Dresden 2	Commonwealth Edison	1970	809	46.6
Dresden 3	Commonwealth Edison	1971	809	52.5
Duane Arnold	Iowa EL&P	1974	550	49
FitzPatrick	Power Auth. State of N.Y.	1975	821	55.2
Ft. Calhoun	Omaha Public Power Dist.	1973	457	59.1
Ft. St. Vrain	P. S. of Colorado[b]	1974	330	—
Ginna	Rochester G&E	1970	490	65
Hatch 1	Georgia Power	1974	822	41.7
Humboldt Bay	Pacific G&E	1963	68	62.2
Indian Point 1	Con Edison	1962	265	—
Indian Point 2	Con Edison	1973	1,033	53.6
Indian Point 3	Con Edison/PASNY[b]	1975	965	—
Kewaunee	Wisconsin P.S.	1974	540	66.7
LaCrosse	Dairyland Power Co-op	1971	50	50.9
Maine Yankee	Maine Yankee Power	1972	855	62.7
Millstone 1	Northeast Nuclear Energy	1971	652	57.7
Millstone 2	Northeast Nuclear Energy	1975	795	58.1

Monticello	Northern States Power	1971	545	67.8
Nine Mile Point 1	Niagara Mohawk Power	1969	620	56.1
Oconee 1	Duke Power	1973	886	57.9
Oconee 2	Duke Power	1974	886	57.5
Oconee 3	Duke Power	1974	886	66.9
Oyster Creek	Jersey Central P&L	1961	640	69.3
Palisades	Consumer Power	1971	880	31.5
Peach Bottom 2	Philadelphia Electric	1974	1,065	59.0
Peach Bottom 3	Philadelphia Electric	1974	1,065	59.2
Pilgrim 1	Boston Edison	1972	655	45.6
Point Beach 1	Wisconsin Electric Power	1970	497	72.6
Point Beach 2	Wisconsin Electric Power	1972	497	71.3
Prairie Island 1	Northern States Power	1973	550	57.0
Prairie Island 2	Northern States Power	1974	550	67.5
Quad Cities 1	Commonwealth Edison	1972	809	—
Quad Cities 2	Commonwealth Edison	1972	809	—
Rancho Seco	Sacramento Mun. Ut. Dist.	1975	913	20.3
Robinson 2	Carolina P&L	1971	700	73.5
St. Lucie 1	Florida P & L	1975	810	16.7
Salem 1	Public Service E & G[b]	1977	1,090	—
San Onofre 1	Southern California Edison	1968	430	73.4
Surry 1	Virginia E&P	1972	820	57.0
Surry 2	Virginia E&P	1973	820	59.5
Three Mi. Is. 1	Metropolitan Edison	1974	830	72.8
Trojan	Portland General Elec.	1976	1,130	9.5
Turkey Point 3	Florida P&L	1972	725	67.6
Turkey Point 4	Florida P&L	1973	725	69.7
Vermont Yankee	Vermont Yankee Power	1972	540	58.5
Yankee (Rowe)	Yankee Atomic Power	1961	175	72.1
Zion 1	Commonwealth Edison	1973	1,100	46.6
Zion 2	Commonwealth Edison	1974	1,100	50.9

[a]The NCR defines "unit capacity factor" as the net power generated, multiplied by 100, divided by the maximum dependable capacity, multiplied by the number of hours in operation.

[b]Plants still in start-up phase of operation.

Source: Nuclear Industry, October 1976, pp. 16–17.

poorer-than-predicted performance as the result of "growing pains." They argue that it is "phenomenal" that the new industry's record is as good as it is, and maintain that reliability will improve as nuclear plants get through the current "learning curve."

Many supporters of nuclear power argue that it has already proved more reliable and cheaper than fossil fuel plants. A press release issued on September 26, 1976, by the Atomic Industrial Forum stated that nuclear power saved an average consumer at Philadelphia Electric $7.92 and one at Green Mountain Power $86 during the first half of 1976.[34]

Commonwealth Edison, the utility that makes the greatest use of nuclear power plants, says that during 1974 its fossil fuel plants operated at a 55 percent capacity factor and its nuclear plants at a 53 percent capacity factor. Byron Lee, a vice president of the company, says the overall costs of generating power were lower for nuclear units: "A review of the actual operating costs indicates that the average cost of our nuclear units was six and one-half mills per kilowatt hour. The same cost for our coal-fired stations burning Illinois coal was nearly nine mills per kilowatt hour."[35] But these figures are often based on a comparison of the fuel costs alone. In discussing savings from nuclear plants, utilities may not include the capital costs of the plant or, if they do, they may make assumptions—as to the capacity at which the plant will operate in the future or as to the life of the plant—which have not been tested.

Nuclear critics argue that the evidence to date does not justify the claims and optimism of the industry's supporters. The Council on Economic Priorities found in a recent study that coal plants 100 megawatts or larger have operated with an average capacity factor of 66.9 percent, compared with 61.6 percent for nuclear plants (450 megawatts or larger) powered by a pressurized-water reactor and 59.3 percent for those using boiling-water reactors. If it were assumed that the coal plants were all base-loaded—some are not, whereas virtually all nuclear plants are—the council said, the average capacity factor for coal plants jumped to 74.8 percent.[36]

Moreover, following up on an analysis done first by David Comey of Business and Professional People for the Public Interest,[37] CEP challenged industry's statements about nuclear power's learning curve. It said there is some evidence that pressurized-water reactors have improved with age, but there is no such trend with boiling-water reactors. And the evidence for pressurized-water reactors above 800 megawatts, or close to the size now being built, it extremely limited. The CEP concluded that "nuclear technology has not developed a capacity factor 'learning curve' in that recently installed units have not operated at higher capacity factors than comparable earlier units at similar stages of maturity."[38] The size of plants being built has increased, and the CEP found that as the size grew, the average capacity factor has decreased by just over 3 percent for every 100 megawatt increase in size. The study attributed lost capacity at the nuclear plants as follows:[39]

	Percent
Scheduled outages (for refueling and routine maintenance)	44
Equipment malfunctions	43
Regulatory restrictions	10
Deliberate reductions because of limited demand	2
(totals less than 100 percent due to rounding)	

The CEP study found that the capacity factor for coal plants also decreased with size—about 1.1 percent for every 100 megawatts of size at the supercritical, high-pressure plants; and 2.2 percent for subcritical plants above 400 megawatts. Coal plants declined at a rate of about 1 percent per year after ten years of operation; the performance of three nuclear plants surveyed declined more rapidly after ten years than did that of coal plants, but the sample of large nuclear plants of that age is too small to allow for accurate prediction of a trend.[40] The CEP study also stated that the operation of coal plants was affected by type of coal burned and estimated that the installation of stack-gas scrubbers could drop capacity performance of new coal-fired units by 3.5 to 5 percent.[41]

Based on its own capacity factor projections—rather than those of industry—CEP concludes that the price of coal may be as much as 30 percent lower than the price of nuclear power, depending on the type of reactor used, even in the northeastern United States where most observers had conceded that nuclear power had an indisputable economic advantage.[42]

The debate over capacity factors is likely to continue until there is more evidence from operation of nuclear plants and large coal-fired plants. Whether a learning curve for nuclear plants exists, as industry argues, after which capacity factors will increase, or whether plants will encounter unforeseen delays with age, as David Comey argues, are among the critical uncertainties plaguing persons who are trying to sort out energy alternatives. As the next chapters show, at one time such questions seemed less significant, there were fewer doubts about the technology, and the future of nuclear power seemed more hopeful.

NOTES

1. Atomic Industrial Forum, INFO News Release, "Nuclear Energy sets New Power Production, Savings Marks in First Half" (Washington, D.C., September 23, 1976), p. 1.
2. Edison Electric Institute, *Information Service News,* "Nuclear Economics" (Washington, D.C., April 6, 1976), p. 1.
3. Donald Cook, "Economics of Nuclear vs. Coal-Fired Generation Considered," *Energy Digest,* September 17, 1975, p. 283.

4. Charles Komanoff, "Responding to Con Edison: An Analysis of the 1974 Costs of Indian Point and Alternatives" (New York: Council on Economic Priorities, August 1975).

5. Ron Lanoue, *Nuclear Plants: The More They Build, the More You Pay* (Washington, D.C.: Center for Study of Responsive Law, 1976), p. 17.

6. Energy Research and Development Administration, "The Economics of Nuclear Power," EDM-068 (3–76), Washington, D.C., 1976.

7. William Rosenberg, "Financial Problems of the Electrical Utilities," Hearings before the Committee on Interior and Insular Affairs, U.S. Senate, August 7 and 9, 1974, pp. 309–10.

8. Charles R. Pierce, in ibid., p. 246.

9. Idaho Nuclear Energy Commission, Gene P. Rutledge, "Analysis of the Economics of Coal versus Nuclear for a Power Plant Near Boise, Idaho," (Idaho Falls: Office of Nuclear Energy Development, March 1976), p. 2.

10. Irwin C. Bupp, Jr. and Robert Trietel, "The Economics of Nuclear Power, 'De Omnibus Dubitandum' " unpublished third draft, MIT, February 16, 1976, p. 1.

11. T. H. Lee, "Power Generation Economics," testimony before the Connecticut Public Utilities Control Authority, January 22, 1976, p. 5.

12. "Huge Plants Demise Signals Trouble Ahead for Energy Expansion," *Wall Street Journal,* September 7, 1976, p. 1. The Idaho Nuclear Energy Commission said in March 1976 that it requires about "11.3 years from the time a nuclear power plant is announced until it attains commercial operation." Idaho Nuclear Energy Commission, op. cit., p. viii. Delays in licensing can cost utilities as much as $5 million per month. "Jersey Utilities Sees Higher Costs in Delay of Nuclear Plant Permit," New York *Times,* September 9, 1976.

13. Sam Ruggeri, "Stack Gas Scrubbing—The 'Con' Viewpoint," *Combustion,* October 1975, p. 16.

14. Federal Power Commission figures, telephone conversation with ERDA officials, October 19, 1976.

15. Charles Komanoff, "Economic Analysis of the Nuclear Expansion Program of Public Service Electric and Gas Co.," testimony before the New Jersey Public Utilities Commission, June 7, 1976, p. 7 (published by the Council on Economic Priorities, New York).

16. FPC figures, telephone conversation with ERDA official, October 19, 1976.

17. Komanoff, "Economic Analysis of the Nuclear Expansion Program of Public Service Electric and Gas Co.," op. cit., pp. 23, 33.

18. W. W. Brandfon, "The Economics of Power Generation," Sargent and Lundy, presented before the Atomic Industrial Forum, Chicago, June 24, 1976, p. 8.

19. Energy Research and Development Administration, op. cit.

20. Brandfon, op. cit., p. 10.

21. Raymond L. Dickeman, "Considerations Relating to the Commercial Viability of the Nuclear Fuel Cycle," presentation to the Atomic Industrial Forum's 1975 annual conference, November 18, 1975, p. 8.

22. William M. Carley, "Uranium Drain," *Wall Street Journal,* June 7, 1976, p. 1.

23. Ibid.

24. Jim Harding, testimony in LILCO, Jamesport, Long Island Nuclear Station Hearings, responses to interrogatories, 1976; see also Michael Rieber, testimony before the State of Connecticut Public Utilities Control Authority, February 10, 1976, p. 20.

25. Federal Energy Resources Council, "Uranium Reserves, Resources, and Production," Washington, D. C., June 15, 1976, pp. 2, 3.

26. ERDA, "A National Plan for Energy Research, Development and Demonstration; Creating Energy Choices for the Future, 1976," Vol. I (Washington, D.C.: April 15, 1976), p. 37.

27. Comments to Investor Responsibility Research Center, 1975.

28. Idaho Nuclear Energy Commission, op. cit., p. 9; Federal Energy Research and Development Administration, "The Economic Impact of Proposition 15," San Francisco, February 20,

1976, p. 20; David Burnham, "Federal Study Charges Little Concern by Utilities in Reliability of Reactors," New York *Times,* March 9, 1975, p. 42.

29. Edison Electric Institute, "Report on Equipment Availability for the 20 Years 1965–74," EE1 75–50, New York, November 1975.

30. Atomic Industrial Forum, op. cit.

31. "U.S. Nuclear Plant Operating Experience," reported in *Nuclear Industry,* October 1976, pp. 16, 17.

32. Commonwealth Edison, "Commonwealth Edison's Nuclear Power Reliability," White Paper, 1975, p. 2.

33. Charles Komanoff, "Economic Analysis of the Nuclear Expansion Program of Public Service Electric and Gas Co.," op. cit., pp. 43, 44.

34. Atomic Industrial Forum, op. cit.

35. Byron Lee, Commonwealth Edison, remarks at the Nuclear Power Plant Availability Workshop, Williamsburg, Va., October 7–9, 1974.

36. Charles Komanoff, *Power Plant Performance* (New York: Council on Economic Priorities, 1976), p. 1 ff.

37. David Comey, "Nuclear Power Plant Reliability: The 1973–4 Record," *Not Man Apart,* April 1975, pp. 12–13.

38. Komanoff, *Power Plant Performance,* op. cit., p. 1.

39. Ibid., p. 37.

40. Ibid., p. 4.

41. Ibid., p. 6.

42. Ibid., p. 7.

CHAPTER

4

THE GOVERNMENT'S ROLE IN NUCLEAR POWER DEVELOPMENT

Both Congress—chiefly through the Joint Committee on Atomic Energy—and the executive branch of government—through the Atomic Energy Commission and one of its successor agencies, the Energy Research and Development Administration—have committed the United States to development of nuclear power as a major source of energy. Critics of nuclear power raise a central question in this regard: Have the political process by which the commitment to nuclear power has been made and the regulatory process by which it has been implemented allowed adequate opportunities for review of alternatives and public participation in decision making?

The dispute about the political and regulatory processes of the government is of interest for several reasons:

Because many of the issues relating to nuclear power involve matters of public policy, members of the public may be inclined to leave resolution of them to the government. But if they conclude that government is not functioning in an appropriate manner, they may wish to increase their involvement and assume greater responsibility.

The regulatory process can have a critical impact on business performance and industry economics. An understanding of the process therefore is useful to an assessment of how particular actions and developments may affect industry.

Major changes in the regulatory agencies and the congressional committees charged with overseeing nuclear development have begun to occur. Many of these may have direct financial and operational impact on the industry. An understanding of the past performance of those agencies and committees, and of concerns about their performance, is useful in understanding and interpreting the significance of the anticipated changes.

HOW THE COMMITMENT WAS MADE

Tradition of Secrecy

The nature of nuclear fission and the unique history of its early development have greatly influenced the process by which decisions relating to the civilian nuclear industry have been made. In the early years all decisions were made in an environment dominated by national security concerns. The tradition of secrecy and the complexity of nuclear technology kept policy decisions in the hands of a few scientists, lawyers, and legislators who had expertise in the area. The Joint Committee on Atomic Energy in Congress and the Atomic Energy Commission on the executive side became the major sources of this expertise and have dominated nuclear policy decision making.

Creation of the AEC

After Hiroshima, U.S. government officials were greatly impressed by the awesome power and potential of nuclear energy. They sought to develop a program that would both preserve U.S. supremacy in nuclear military technology and allow nuclear power to be developed for peaceful purposes. In 1946 Congress adopted legislation, recommended by President Truman, to transfer responsibility for nuclear development from the military to a five-member civilian Atomic Energy Commission whose purpose was "government control of the production, ownership and use of fissionable material to assure the common defense and security and to ensure the broadest possible exploitation" of nuclear technology. The commission was put in charge of the vast existing nuclear industrial complex, together with the research and development program that had begun during World War II in support of the Manhattan Project.

Creation of the Joint Committee

Concurrently, Congress created the Joint Committee on Atomic Energy to oversee the activities of the AEC and gave it unique powers: it became the only standing joint committee of Congress with powers to recommend legislative action and to serve as the conference committee to reconcile any differences in legislation within its jurisdiction that was passed by both houses of Congress. (In general, on other kinds of legislation, the conference committees include both members of the standing committees that support the legislation and other members of Congress who have opposed the legislation or, more often, who have sponsored significant amendments on the floor.) The creation of the Joint Committee was Congress's answer to the problem of maintaining legislative control over a highly technical program, a major portion of which was to be conducted under conditions of secrecy.[1]

As its interest and experience grew, the Joint Committee expanded its oversight role to assume more direct responsibility for the civilian nuclear program. The AEC was somewhat preoccupied with cold war demands for development of military programs, according to Craig Hosmer, a former member of the Joint Committee, and the Committee was developing a good deal of expertise concerning civilian nuclear technology. The Committee came to play the primary role in what Hosmer describes as a "partnership" relationship between the AEC and the Committee. While newly appointed AEC commissioners often knew little about nuclear power, six long-time members of the Joint Committee became very knowledgeable about nuclear power and, consequently, very influential. Hosmer recalls: "It was one instance in which power flowed to a congressional committee from the executive, rather than the other way around."[2] By 1957 the Joint Committee had assumed responsibility for policy development and programming, and the AEC was involved primarily with the details of program administration. The Joint Committee established for itself certain executive rights: in 1961 the Committee's chairman wrote that the Committee "reserves the right to recommend projects and levels of support which it believes necessary or important to national interests."[3] The dominant role of the Joint Committee, until recently, went unchallenged. With one minor exception, before 1974 no bill proposed by the Joint Committee was questioned on the floor of Congress.

Encouragement of Private Industry

Early on, the Joint Committee and certain members of the executive branch began to consider how to shift some of the responsibility for civilian nuclear power development to private industry. By 1954 the Committee considered nuclear power to be at the threshold of commercial viability and began to explore methods of transferring at least a portion of the government's full control to private companies. The Committee drafted the Atomic Energy Act to authorize the AEC to license private parties to work on all kinds of nuclear materials except weapons materials. The legislation was enacted in 1954.

Attracting private participation was not easy. Consequently, the government adopted a number of programs to encourage utilities to invest in nuclear power plants. Among these were government funding of research, development, and demonstration projects and a system of federally subsidized insurance and of limited liability in case of a nuclear power plant accident.

The Power Demonstration Reactor Program

Utilities were reluctant to commit large amounts of capital to nuclear technology, which they viewed as untested and not fully developed. To spur investment and demonstrate the commercial viability of nuclear power, the Power Demonstration Reactor Program was launched in 1955.

Under the program, the AEC made a major commitment to research at government laboratories on problems vital to development of a civilian nuclear industry, financed research on reactors that were more sophisticated than the first generation of light-water reactors, paid the costs of manufacturing reactor cores, and supplied nuclear fuel free of charge. The program met with considerable technical success, highlighted by the successful operation of a commercial nuclear plant at Shippingport, Pa., under the direction of Admiral Hyman G. Rickover. This led to considerably increased interest in nuclear power by manufacturers and utilities.

Price-Anderson Act

Both manufacturers and utilities remained concerned about their potential liability if a major accident should occur at a nuclear power plant. A study of potential risks and damages commissioned by the AEC in 1957 made private companies even more apprehensive than they had been. General Electric made plain that it would withdraw from major participation in nuclear development unless the government came up with a program limiting the company's potential liability. "It had become clear," according to George Washington Law School professor Harold Green, "that private enterprise interest in nuclear power would dissipate almost entirely unless some formula were found to enable private industry to participate without risking public liability."[4] Because of the extremely limited experience that had been obtained with reactors, insurance companies were not willing to provide more than $60 million in liability insurance.

Responding to these concerns, in 1957 the Joint Committee developed and secured passage of what is known as the Price-Anderson Act. The act limits the liability of an individual company and provides government subsidies to cover liabilities well beyond those that private commercial insurers were —or are now—willing to cover. The act served the dual purpose of facilitating public compensation in the case of a nuclear power plant accident and eliminating a major barrier to the commercial development of nuclear power—easing businesses' fears about the potentially bankrupting liability that could result from a nuclear power plant disaster. It was named for its sponsors, Representative Melvin Price (D-Ill.) and Senator Clinton P. Anderson (D-N.M.).

The Price-Anderson Act limits overall liability for losses incurred as a result of a nuclear power accident to $560 million. A portion of the liability is assumed by private insurance purchased by utilities from a common fund. Utilities are obligated to purchase as much insurance as private insurers are willing to provide—$140 million today. The remainder—now $420 million—is covered by a government obligation to subsidize in the event of a nuclear power plant accident.

In December 1975 Congress amended the act and extended it until 1987. The most important amendment provides for a reduction in the government's obligation as the nuclear industry grows. For the moment, the overall liability ceiling remains at $560 million. But to finance that amount, in case of a nuclear accident, all utilities using nuclear power will be asked to pay a "deferred premium" of between $2 million and $5 million for each nuclear plant they operate. These premiums, combined with the contributions of private insurers, are expected to cover an increasing portion of the liability, and the federal government will continue to subsidize the rest, up to the ceiling of $560 million.

Once the combined insurance company/utility contribution reaches $560 million, the liability ceiling will be lifted, and the liability coverage will be allowed to grow above $560 million as the combined contribution increases.

In addition, it is possible that Congress will be asked to pass emergency legislation if claims in excess of $560 million are made. Under the act, all losses would be repaid irrespective of fault by the utility, manufacturer, or architect. This provision, added in the 1960s, makes recovery far easier than it would be if the claimant had to prove all facts under the law of a particular state. Insurance companies participating in the insurance pool established under the Price-Anderson Act say that the operating performance of nuclear plants has indicated that original premiums charged to contributors were too high. To date, only a few small claims have been made, and no government payments have been required. As a result, in the first nine years of operation, the insurance companies returned to contributors about 70 percent—$8.3 million—of the premiums received.[5]

Other Incentives

In general, where industry has been hesitant to commit itself, the AEC (and now the Energy Research and Development Administration) has spent money to demonstrate that civilian nuclear power is feasible. In several areas—safety research, safeguards against damage from nuclear plants, enrichment of fuel for nuclear plants, storage of fuel, and disposal of radioactive wastes—ERDA continues to dominate.

THE REGULATORY PROCESS

Until late 1974—when the functions of the AEC were split between the newly formed Nuclear Regulatory Commission (NRC) and the Energy Research and Development Administration—the AEC was both the prime supporter and the chief regulator of the nuclear industry. Harold Green described the AEC as "by far the largest entrepreneur in the industry, the largest consumer of the industry's materials and services; it plays an active role in promoting the industry and in encouraging and subsidizing private interests to enter the industry at the same time that it is a potential competitor of these interests; and, finally, it licenses and regulates the private firms which it has encouraged and subsidized."[6]

The regulatory role of the AEC and its successor agency has developed substantially in recent years. As John O'Leary, former AEC director of licensing, told the Investor Responsibility Research Center (IRRC) in 1974, "The industry was only a penny industry until the early 1970s."[7] There was little system to the regulatory process; AEC engineers made ad hoc judgments about the adequacy of individual nuclear plants, O'Leary said. Until 1971, no serious efforts were made to develop standards, guidelines, and routine procedures for plant construction as part of the regulatory process. According to the AEC's director of regulation, L. Manning Muntzing, in the 1960s, "at the AEC, the standards zealots could meet in a phone booth."[8] Eugene Cramer, an official at Southern California Edison Company, explains that "until there were substantial sales of multiple units of essentially similar design at about the same time, by the same vendor, the conditions for development of standards did not exist."[9] NRC Commissioner Victor Gilinsky told the Joint Committee on Atomic Energy in March 1975 that "the nuclear regulatory system requirements in a sense grew by accretion. Responsible persons decided that more and more requirements were useful."[10]

This situation changed rapidly in the early 1970s, as the number of applications for nuclear power plant licenses increased dramatically. According to O'Leary, after 1971 "the regulatory staff expanded from the dozens into the hundreds." The AEC also moved to establish standards and uniform criteria to be used in the licensing process. In 1972, a full-time staff charged with developing standards was organized.

Development of the AEC's regulatory processes was further stimulated in 1971 by the decision of the U.S. Court of Appeals for the District of Columbia in *Calvert Cliffs Coordinating Committee* v. *AEC.* The court ruled that the AEC was obliged under the National Environmental Policy Act to make a detailed assessment of the costs, benefits, and environmental impact of nuclear power plants before it issued licenses for them.

Following the establishment of the NRC, there was a seven-month lull in the issuance of permits and licenses, during which time the new Commission undertook a major organizational effort. This effort brought an expansion of the regulatory staff, the introduction of a number of new standards and guides, and the launching of several studies to develop means of streamlining the licensing process. The results of these efforts—and the effectiveness of the regulatory program in general—are the subject of some debate among the supporters and critics of nuclear power.

Development of Standards

The various standards affecting nuclear plants—ranging from the technical specifications governing component parts to generic standards for radioactive emissions levels or seismic considerations in plant siting—are derived from a number of sources. Some may have existed in engineering codes. Others have evolved through ad hoc decisions made by NRC engineers reviewing individual plant designs. In many areas, industry or academic groups have made suggestions which are issued as criteria in regulatory guidelines if the NRC accepts them. Rule-making hearings are held to establish standards on generic issues such as criteria for emergency core-cooling systems, low-level radiation emissions, environmental impact of the uranium fuel cycle, and the impact of radioactive materials transport.

Since 1971, AEC and NRC staffs have issued more than 300 "Regulations and Guides" establishing acceptable levels of risk or endorsing nearly 200 national standards developed through the volunteer standards program.[11] Under the program, approximately 8,000 technical specialists, primarily from industry, have served on standards-writing committees. Some industry and NRC representatives see the end in sight. O'Leary said in 1974, "The past year has been encouraging. It looks like the technology may be maturing enough for a long period of stability in regulatory requirements. We have a good deal of confidence that there is not too much wrong with the basic technology. We couldn't say that a year ago."[12] In 1975 the NRC completed a set of "standard review plans," covering more than 1,400 pages, to be used by the NRC technical staff in reviewing nuclear power plants. Some industry officials expect the plans to freeze regulatory requirements. The NRC says only that the plans will improve "over-all quality, uniformity and predictability of staff reviews" and will serve as a "well-defined basis for evaluating proposed changes."[13]

Licensing Procedures

In order to build a nuclear power plant, a utility must follow a number of procedures, including participation in two public hearings and numerous reviews by federal and state agencies. To obtain a license from the NRC, it must be able to show, in the language of the Atomic Engery Act, that "there is reasonable assurance that the proposed facility can be constructed and operated at the proposed location without undue risk to the health and safety of the public."

The NRC requires applicants to use a "defense-in-depth" approach. Plant components should be designed for maximum safety in normal operation and maximum tolerance if the system malfunctions. Utilities are required to assume both that accidents will occur in spite of care in design, construction, and operation, and that some protective systems may fail simultaneously. Nuclear plants must incorporate features emphasizing "quality," "redundancy," "inspectability," and "testability"; they must have safety systems to protect operators and the public in the case of accidents, including multiple safety systems where appropriate.

The nuclear power plant licensing process has five basic steps (see figure 3).

Application

The utility submits to the NRC a formal application for a construction permit. The application must include general information to assure that the utility has the financial capacity to build and operate a reactor; a preliminary safety analysis of the reactor design and site, detailing aspects of operations, personnel, accident procedures, radioactive releases, waste storage, and quality assurance programs; an environmental report which becomes the basis of the NRC's environmental impact statement and sets forth probable impact, adverse environmental effects, and alternatives; and information relevant to antitrust considerations. A nonrefundable fee of $125,000 must be paid with the application.

Review by NRC Staff

If the application is accepted, it is passed along a three-track review system covering environmental, safety, and antitrust considerations. For the environmental review, first the staff examines the utility's environmental report; then the staff drafts an environmental impact statement and sends it to a number of federal, state, and local agencies in compliance with the National Environmental Policy Act; then it sends a final statement incorporating any agency comments to the Atomic Safety and Licensing Board—an independent review board drawn from a panel of lawyers, economists, and scientists—for approval.

For the safety review, the staff reviews the preliminary safety assessment report and discusses it with utility and outside representatives. After review the staff issues a safety evaluation report which is passed to the Advisory Committee on Reactor Safeguards—an independent board whose members, mostly scientists and engineers, are appointed to four-year terms by the commission. After it reviews the staff's report, the advisory committee writes a letter to the Nuclear Regulatory Commission, and the staff submits a supplemental safety evaluation report reflecting the advisory committee's findings.

The antitrust review is conducted by the Justice Department. According to former Assistant Attorney General Thomas Kauper, the aim of the department is to ensure that utilities are making efforts "to promote efficiency, and, as part of that, to save scarce fuel. Coordination and reserve sharing of the type we have been seeking in the (NRC) proceedings enables all utilities to use large efficient units, units which produce low-cost power with less fuel."[14] The Justice Department issues an advisory opinion to the Atomic Safety and Licensing Board.

Atomic Safety and Licensing Board Hearings

When it receives completed reports from the NRC and the Justice Department, the Atomic Safety and Licensing Board holds public hearings to review any remaining issues. The hearings in the three areas—environment, safety, and antitrust—may be held independently or simultaneously. Usually the environmental report is completed before the preliminary safety report and, because the utility is anxious to obtain a limited authorization to begin construction, the environmental hearings are held first. The hearings are quasi-adjudicatory. Formal notice is provided and a prehearing conference is held to consider petitions from parties wishing to intervene in the proceedings. Intervenors—those who can show they will be affected by a proposed nuclear plant—are granted the right to offer evidence and to subpoena documents and witnesses and to cross-examine witnesses at the hearings. Hearings are limited to technical and environmental issues relating to the specific application; standards, regulatory procedures, and general policy questions are not reviewed. After completion of the hearings, the board decides whether to accept, conditionally accept, or reject the application.

Review of the Licensing Decision

Any person participating in the hearings as a party or an intervenor may appeal a board ruling to an Atomic Licensing Safety and Appeal Board. A further appeal may be made to the federal courts, although most observers consider it unlikely that the courts would reverse an appeals board's administrative decision.

Operating License

Once the utility has received its construction permit and completed construction, it must go through a second set of reviews and hearings, in order to obtain its operating license. The utility must file a final safety analysis report and an updated environmental impact statement. Both are reviewed by the NRC staff, and reports are filed with the Atomic Safety and Licensing Board. Further hearings at this stage are not required and are held only if an interested party achieves the status of intervenor. The board's decision on an operating license is subject to the same rights of appeal as is the construction permit. Once its operating license is approved, the utility must pay a $250,000 licensing fee and a second fee based on reactor size.

FIGURE 3
Parallel Tracks in Construction Permit Review Process

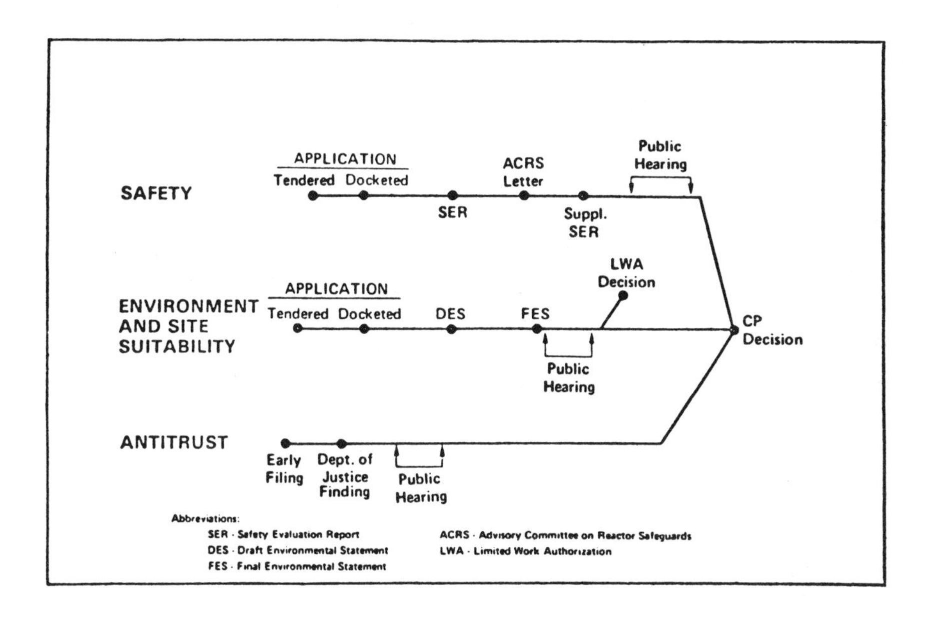

Source: U.S. Nuclear Regulatory Commission, 1975 Annual Report, p. 14.

Proposals to Speed Licensing

The existing procedures for licensing require about 11 years from the decision to build to the beginning of operation of a nuclear plant. Both industry and the NRC have made suggestions on how to speed the licensing process. Hearings would be limited to review of the applicant's ability to meet technical standards. Extraneous or generic issues—often controversial and time-consuming to resolve—would be handled in rule-making hearings held to establish general policy and industrywide standards. (Rule-making hearings are administrative, rather than adjudicatory, and thereby permit administrative discretion in limiting intervenors' opportunity to present evidence, cross-examine witnesses, or subpoena information.) To the extent possible, plant design would be standardized, so that once a basic plant design is approved, similar design plans for individual plants would not require extensive review. The NRC estimates that by standardizing plants and by approving potential power plant sites at some time before the decision is made to build a plant on the site, the time between a decision to go nuclear and start-up of plant operations can be reduced from 11 to 6 years.

Enforcement

The NRC has nearly 400 staff members assigned to inspect the nuclear plants now operating or under construction. NRC officials told IRRC that a plant is inspected an average of 50 times during the period before operation and must be inspected a minimum of four times a year after it goes into operation. Any violations found during the inspections are reported orally to the plant manager, and the NRC staff writes a letter to the management of the utility asking that the problem be rectified. If necessary, the NRC will order the plant to operate at less than capacity or to shut down until the problem can be corrected. The NRC issues an order to shut down or imposes civil fines only after repeated violations have indicated what the NRC considers "a pattern of noncompliance." The NRC argues that, particularly with power plants, civil penalties are unnecessary for the most part. "The greatest penalty," one official said, "is to require the plant to shut down, forcing it to buy replacement power (often at a cost of $100,000 to $200,000 per day) elsewhere. A civil penalty's largest cost—the NRC is limited to a $5,000-per-violation ceiling per 30 days—is the stigma attached to it."[15]

During the period between July 1, 1975 and September 30, 1976 the NRC listed 1,611 items of noncompliance. Only six of these were considered serious violations, 923 were classified as infractions, and 682 were noted as deficiencies. The NRC issued fines to ten utilities totaling $172,250 between July 1, 1975 and December 15, 1976. NRC officials report that the limited use of fines and the efforts to get industry to regulate itself have worked. "By and large," one NRC official told IRRC, "I think our enforcement program is effective."[16]

THE CONTROVERSY OVER GOVERNMENT'S ROLE

Industry and the NRC view the political and regulatory processes as a sincere effort to protect the public interest while allowing for the necessary growth of nuclear power. Both company and governmental officials are critical, however, of what they see as inefficiencies in the regulatory process. Critics see the processes as detrimental to the public interest, providing few opportunities for genuine consideration of the concerns of nuclear power opponents, and emphasizing the promotion of nuclear development over its regulation.

Views of Nuclear Power Proponents

Industry and government officials dispute contentions by critics that the decision to develop nuclear energy was made without sufficient public involvement. They say that the current process of approval allows for public review of nuclear power and argue that both industry and government have made an effort to provide the public with information on issues relating to nuclear development. They also argue that the relationship between industry and government has been healthy and that government assistance to companies involved in aspects of nuclear power has not interfered with its regulatory responsibilities. If anything, industry officials say, the government has been too strict in its regulation.

Public Involvement in Decisions Relating to Nuclear Power

Government and company representatives say that the decision to proceed with the development of nuclear power was made only after conscious deliberation by public representatives. "The fact is," Westinghouse wrote to IRRC, "that the entire Congress enacted the legislation—the Atomic Energy Act. It is specious to argue that the decision to go nuclear was not adequately considered."[17]

Industry spokesmen agree with critics that the licensing process does not provide a forum for the discussion of broad public policy issues relating to nuclear power development. Those issues they consider to be political. "They were never meant to be considered in the licensing process," one representative told IRRC. "They belong in a political forum."

But the process does allow for public participation in the review of ongoing decisions relating to nuclear development, according to government representatives. Then-NRC Commissioner William Anders once said, "We welcome public participation in our decision-making process . . . and . . . our own process is one that does provide ample public participation and does not inhibit dissent within the staff."[18]

Availability of Information

The AEC in 1974, and later the NRC, recognized that past restrictions on disclosure of information to the public had been overly harsh—even suppressive—but officials say the policy has changed. Muntzing of the AEC said in 1974 that "three years ago we created a revolutionary openness—we may not be perfect, but we're a lot better." He told the New York *Times* that "there is no agency as dedicated to opening up as the AEC."[19] Then-NRC Chairman Anders testified in 1976 that "an examination of our record will show further that this commission has gone well beyond the requirements of the Freedom of Information Act in making information available to the public. . . . The NRC is committed to giving the public the facts about nuclear safety, whether the facts are agreeable or not. . . . Every document representing an official staff position on a reactor safety issue is routinely placed in our central public document room as well as in other locations we have established around the country." He commented that "the public may add new insights to this information and can provide us with valuable feedback, by letter or in public hearings, which we consider in reaching our determinations."[20] Marcus Rowden, who has succeeded Anders as NRC chairman, has commented: "It is just not in our own self-interest to cover up or appear to cover up."[21]

Internal Dissent

In response to allegations by critics that the NRC does not allow sufficient internal dissent, Chairman Rowden stated in November 1976: "The commission is intent that there be no misunderstanding our commitment to these basic principles—that staff be able and expected to make known their best professional judgment, whether or not it corresponds with the views of other staff or management; and that this can be done with the assurance of no recrimination or retribution."[22] The NRC issued a series of memos in November 1976 to assure employees that any safety concerns they may have will be examined by the commission.[23]

Relationship with Industry

Company officials recognize that the government has played an active role in the development of nuclear power, but they argue that government assistance is fully justified. Richard McCormack, president of General Atomic, describes the industry as "born in government and consciously weaned by statutory and administrative policies." Such involvement was, and continues to be, necessary because, he says, the nuclear industry

> is the most technologically intensive major business in the world today. It requires enormous amounts of front-end capital to bring a system into full commercialization. . . . Traditional short-time manufacturer-customer commercial practices are breaking down because they involve time-spans and risks that simply do not allow business to proceed on an equitable basis. . . . The private sector first isn't investing in reprocessing and recycle because, frankly, the vendors find themselves against the wall with neither the capital nor, sadly, the desire to make new investments in an industry absolutely essential to the growth of our nation, but beset with uncertainties no pruduent businessman would accept.[24]

The answer, McCormack says, is that this country must realize that the "commercialization of the nuclear industry is a task . . . that is new and without precedents and, frankly, a whole lot tougher than anyone realized."

Most industry representatives agree and say that more, rather than less, government involvement will be required in the future to assist in the development of methods for waste disposal, reprocessing spent fuels, and new technologies for enrichment of uranium. Greater expenditure on nuclear power than on alternative technologies or conservation is justified, industry sources say, because "the costs of research increase sharply as you move up the learning curve." General Electric representatives commented to IRRC that "to compare research funds for an advanced technology with those allocated to a new technology is meaningless."[25]

But nuclear supporters deny that government participation in the development of the industry has affected the AEC's or the NRC's ability to regulate it, and they reject charges that either may have colluded with industry to spur development. Then-Chairman Anders testified in 1976 that "our overriding goal and consideration is safety, and though we are interested in regulatory efficiency, we will take as long as necessary to ensure the plant is safe before it is allowed to operate."[26] Charges of collusion, Romano Salvatori of Westinghouse said in 1975, are "totally false and without foundation." He asked, "Is there conspiracy [between the AEC and industry] when on more than one occasion the AEC has failed to grant construction permits for construction . . . has held up licensing of already constructed reactors . . . when public hearings proceed over many months or even years . . . when the AEC requires design modifications, sometimes costing many millions of dollars?"[27] And NRC Chairman Rowden wrote to Ralph Nader in 1976: "The strong criticism we have received from the regulated industry, responding to what it views as undue regulatory conservatism, reflects the reality that NRC has taken measures it deemed necessary notwithstanding substantial impact on the industry."[28]

In fact, many company officials see industry as the chief victim of the regulatory process. George J. Stathakis of General Electric describes the process as "extremely rigorous . . . difficult and expensive." He says that GE's "relations with the regulatory staff had been an adversary character. We have certainly not found the NRC to be overly impressed by economic considerations. From our perspective we have often felt that the NRC has paid excessive deference to the concerns expressed by the opponents of nuclear power and has acted in an overly conservative manner."[29]

A major complaint has related to the standards set by the NRC and by the AEC for plant construction and operation. The absence of standards has slowed approval of plant design in certain areas, and overly strict standards have raised construction costs in others. Although many industry representatives have played a role in developing standards, one utility president wrote to the commission in 1974 that unless a more realistic approach is taken to standards, "there will be a marked falling off of industry enthusiasm to devote so much time and effort to standards development." Two major problems, according to McCormack, are ratcheting—the "practice of continuously escalating regulatory requirements on an ad hoc basis from one project to the next"—and backfitting—"requiring redesign of facilities to conform to subsequently adopted regulatory requirements." Some industry representatives say that new costs caused by unnecessary conservatism and ratcheting of standards should be absorbed by the government.[30]

Inefficiency in setting standards has been a major cause of expensive delays in the licensing process, according to industry officials. In the last few years, the time span between a decision to apply for a license and the start-up of operation has stretched from 6 to 11 years. Spokesmen attribute the delay to problems in setting standards, to difficulties in making changes in existing or planned equipment to meet new standards, and to the proliferation of actions by intervenors against nuclear plant construction.

Views of Nuclear Power Opponents

Critics of nuclear power have a number of complaints about the role that government has played in its development. Until recently, they say, the growing commitment to nuclear power, has never been openly reviewed in a political forum, and they argue that the current regulatory process offers no opportunity for debate on the major issues involved in its development. They also maintain that the agencies and committees involved in overseeing or administering the nuclear power program—the Joint Committee on Atomic Energy, the Atomic Energy Commission, the Energy Research and Development Administration, and the Nuclear Regulatory Commission—have been biased in its favor. As a result, they say, they have ignored energy alternatives and have failed to provide the public with a balanced assessment of nuclear power.

Joint Committee

Opponents of nuclear power are very critical of the Joint Committee on Atomic Energy. They say the Committee decided quite early that its major role was to promote nuclear power. Thereafter, they say, it has made no efforts to reexamine its initial decision but has concentrated instead on solving technical problems which would impede development of a civilian industry. Ralph Nader, in testimony before the Committee, commented: "It is revealing to examine how poorly the Committee has performed its oversight function with regard to protecting the public against the wide range of dangers associated with nuclear power development," and he accused the committee of actively assisting the AEC "in suppressing critical data on atomic plant safety hazards."[31]

Critics assert that other members of Congress have chosen to defer to members of the Joint Committee on the theory that they possess the experience and technical knowledge required to review nuclear power. Thus, according to Green of George Washington Law School, "the development of atomic energy policy has taken place largely within a closed circle of government atomic energy specialists, on the apparent assumption that atomic energy represents a totally unique and isolated problem separate from other technical concerns of the government."[32] Until recently, this isolation, critics say, made it impossible for questions about nuclear power to get a full hearing in Congress.

AEC

During its lifetime, the AEC was criticized heavily by opponents of nuclear power. They considered the commission to be schizophrenic, with its dual purpose of regulation and promotion, and they argued that its performance was unbalanced in favor of promotion. Anthony Roisman, a lawyer for a number of groups intervening in hearings to oppose nuclear plant licenses, said in 1974 that "promotion has often clouded the AEC's regulatory judgment."[33] Daniel Ford of the Union of Concerned Scientists emphasized this point in December 1974 in a letter to IRRC:

> For the most part, we sense that people in the country generally have thrown up their hands in despair when it comes to any serious discussion of the nuclear power controversy. . . . Instead, they have trusted the AEC to make

> expert impartial decisions on nuclear safety matters in the public interest. In fact, however, the AEC has acted much more to promote nuclear power than they have to protect the public health and safety. . . . The fact that the agency entrusted with making the safety decisions cannot itself be trusted is a factor that must weigh heavily in the country's acceptance or nonacceptance of nuclear power. No one doubts the need for strict controls over nuclear power, and if the agency insuring such strict controls is seriously compromised, then it's a major institutional barrier to the society's use of nuclear power.[34]

In support of their contention that the AEC sacrificed safety in order to speed growth, critics refer to a 1972 article in *Science* magazine by Robert Gillette. Gillette wrote that a number of researchers within AEC laboratories had "become convinced that the AEC in its eagerness to develop a thriving industry . . . has deliberately by-passed tough safety questions still hanging over ordinary, water-cooled reactors."[35] Not only did the AEC overlook safety, the critics say, but the commission also tried to keep information concerning safety problems from the public. David Burnham of the New York *Times,* after examining a number of AEC documents written before 1973, concluded that "AEC documents show that for at least 10 years, the commission has repeatedly sought to suppress studies by its own scientists that found nuclear power was more dangerous than officially acknowledged or that raised questions about reactor safety devices."[36]

ERDA

A number of observers—including most critics of nuclear power—argue that ERDA is dominated by former AEC employees and that the thrust of its programs has been too heavily focused on the development of nuclear power. President-elect Carter commented during the 1976 presidential campaign that "because [ERDA] is an off-shoot of the now defunct Atomic Energy Commission, its entire slant is toward the nuclear industry. Sixty-five percent of its research resources for fiscal year 1977 are oriented toward nuclear fission and fusion, while only 5 percent will go to energy conservation and 6 percent for solar power."[37] Carter described this distribution as "folly," and a number of critics agree with that description. They argue that ERDA must reverse its priorities to expand government research on conservation and alternative energy sources such as solar and fusion power.

NRC

Many of the same complaints raised about the AEC are being renewed by critics of the NRC. Persons concerned about the development of nuclear power comment on the lack of public participation in the decision-making process, the role of government in promoting nuclear power, the inadequacy of government regulation, and the lack of information available to the public from government and industry.

Public Participation in Decision Making

Critics of the Atomic Energy Commission argued that the way it structured hearings—with the intent of reviewing the applicant's capacity to meet AEC standards—did not allow debate of what intervenors saw as the major issues of substance—plant safety, safeguards, and waste storage. The process has changed little under the NRC, and Roisman argues that "issues relevant to the determination of whether nuclear reactors should be built and operated are still excluded from the hearings."[38]

Intervenors assert that limiting licensing discussions to technical issues tilts the process unfairly in favor of industry. The problem, Green, of George Washington, has said, is that intervenors "are driven to contest issue of the license on grounds with respect to which they have the least technical resources and the least competence. The experience is characteristically one of frustration, and the opponents of nuclear power emerge from their inevitable defeat with a feeling of alienation and resentment about the stacked deck of cards."[39] Most intervenors lack the expertise needed to present their case solidly. If they have the funds to pay for expertise, they are often unable to find it. Steven Ebbin and Raphael Kasper wrote in a 1974 National Science Foundation report, *Citizen Groups and the Nuclear Power Controversy,* that "the AEC's role as the major funding source for research in the nuclear sciences over the last 25 years has created a situation in which almost every nuclear engineer of worth is under grant or contract to the AEC or in the employ of the AEC, the nuclear industry or a national laboratory. . . . The level of scientific and technical advice to intervenor groups is marginally adequate, if that."[40]

Critics also charge that the AEC's and now the NRC's policy of discussing generic issues in rule-making hearings further limits the opportunity to challenge nuclear development. They argue that by removing major issues from adjudicatory licensing hearings to administrative rule-making hearings, the NRC potentially inhibits the intervenors' ability to raise questions and to get important information before the public.

Such efforts to speed the licensing process are unnecessary, they say. Critics—such as Amory Lovens of Friends of the Earth—argue that interventions in hearings are not a major cause of delay. Some critics refer to a report by the Atomic Industrial Forum showing that legal challenges ranked seventh among the ten most important causes of delays in construction of 28 nuclear plants scheduled to begin operation in 1973:[41]

Cause	Number of Plants Affected	Plant-Months of Delay
Poor productivity of labor	16	84
Late delivery of major equipment	9	68
Change in regulatory requirements	8	23
Shortage of construction labor	5	21
Strikes of construction labor	5	18
Equipment component failure	6	15
Legal challenges	4	9
Strike of factory labor	4	6
Rescheduling of associated facilities	1	12
Weather	1	5

They argue that the hearings process is one of the few places where they can obtain public attention. According to Roisman, lawyer for several intervenors, although some foes of nuclear power will do anything necessary and legal to stop proliferation of nuclear plants, most see the use of the limited issues that can be raised at hearings as one legitimate way in which to raise public consciousness.[42]

Adequacy of the NRC

Critics of nuclear power say that the commission's regulatory objectivity suffers from conflicts of interest among its staff members, many of whom are drawn from—or will eventually go to—the industry they regulate. Common Cause charged, in a study released late in 1976, that relationships between NRC staff and consultants and the industry it regulates offer serious potential for conflict of interest. The study reported that 72 percent of the NRC's top 429 employees have been employed by private energy companies and that 90 percent of these employees came from companies with which the NRC had current contracts or licenses. Sixty-five percent of the NRC's consultants are working as well for companies that have received NRC licenses or contracts. "Our findings," Common Cause wrote in the study, "point to potential conflicts of interest, and the possibility of serious agency bias, throughout the executive bureaucracy."[43]

The NRC's ability to regulate is also affected by the fact that most of its staff members are drawn from what was the AEC, according to some critics. They say that the NRC continues to be affected with the AEC's promotional bias. In an October 25, 1976 press release, Ralph Nader alleged that the NRC was supporting subsidies worth more than $18 billion to utilities interested in building nuclear plants, and he stated that certain NRC activities violated the spirit and the letter of NRC's responsibilities as a regulatory agency.[44]

In protest of what they consider to be an inadequate response to safety questions, several engineers left the NRC and three quit the nuclear industry in 1976. Robert Pollard, a reactor engineer and project manager with the NRC until February 1976, testified that, "as a result of my work at the commission, I believe that the separation of the Atomic Energy Commission into two agencies has not resolved the conflict between the promotion and regulation of commercial nuclear plants. Because I found that the pressures to maintain schedules and to defer resolution of known safety problems frequently prevailed over reactor safety, I decided I had to resign."[45]

In October 1976, Ronald Fluegge, an NRC engineer, resigned and asserted that he had been repeatedly "frustrated in [his] efforts to make the agency deal honestly with pressing nuclear safety problems."[46] Fluegge stated, "We are issuing safety evaluation reports that are carefully censored to conceal major safety problems. We are withholding from the public NRC staff technical analyses of a wide range of unpleasant nuclear safety difficulties. We are giving the public glib assurances about the nuclear plant safety that we know lack an adequate technical basis."[47] Burnham of the New York *Times* reported in October that nine engineers still employed by the NRC "have charged in recent interviews that the agency has refused to consider and act upon some serious questions they have raised about the safety of nuclear reactors."[48]

The turmoil within the NRC followed the resignation earlier in the year of three engineers at General Electric, who said they left GE because they felt that the nuclear program was being developed too rapidly to allow for adequate testing and analysis of operating procedures.[49] They said the NRC had been "misleading" in its public statements and that it has failed to convey to the public some of the real problems involved in nuclear development.[50]

Court Support

In at least one instance critics' complaints have been supported in the courts. The Natural Resources Defense Council (NRDC) argued successfully that the NRC's licensing process failed to deal adequately with the environmental impact of nuclear power wastes. Ruling in favor of the NRDC and National Consolidated Intervenors against the NRC, the U.S. Court of Appeals for the District of Columbia found in July 1976 that the NRC failed to give adequate attention to the environmental impact of reprocessing and storage of waste materials produced by nuclear power plants. As a result, the NRC temporarily halted the issuance of all licenses, and nuclear opponents petitioned the commission for a reexamination of licenses at plants already operating. The NRC issued a provisional interim rule in October 1976 which would allow it to grant provisional licenses until a final rule could be drawn up following a public-hearings process that could extend for 18 months.[51]

THE FUTURE ROLE OF GOVERNMENT

In the last two years there has been a growing effort to decentralize decisions relating to nuclear power. The AEC has been split, and there has been a de facto expansion of congressional oversight responsibility which has cut deeply into the control once exercised alone by the Joint Committee on Atomic Energy. The inauguration of a Carter administration and changes among committee responsibilities in Congress seemed certain to carry the decentralization well beyond that which had occurred since 1974.

President-elect Carter criticized ERDA for its "slant" toward the nuclear industry. He said he would abolish ERDA and incorporate certain NRC functions into a new cabinet-level department of energy. Carter's reorganization of energy agencies was expected to diminish somewhat the position of nuclear power relative to conservation of energy and to the development of alternative energy sources.[52]

In Congress, by the beginning of 1977, opposition to a break-up of the Joint Committee on Atomic Energy appeared to be on the wane. Three of nine Senate members of the committee lost their seats in the November 1976 elections and two others retired. On the House side, two members, Representatives George Brown (D-Calif.) and Teno Roncalio (D-Wyo.), actually favored dissolution of the Joint Committee, and Mike McCormick (D-Wash.), one of the committee's most active members, appeared to be willing to accept a decentralizing of its responsibilities.[53] The Senate actively considered in 1976 transferring the Joint Committee's major responsibilities to the Interior and Armed Services Committees. The Government Operations Committee assumed a certain oversight role when it held hearings on energy reorganization and on the exports of nuclear materials, and a number of other committees began making jurisdictional claims on portions of the energy program.

Discussion of nuclear power has entered the political forum, and debate over various aspects of its development is likely to exert increasing pressure to slow the speed at which it develops.

NOTES

1. For a history of the early relationship between the nuclear industry and government, see Harold P. Green, "Nuclear Technology and the Fabric of Government," 33 *George Washington Law Review* 121, 1964, pp. 121 ff.; Arthur D. Little, Inc., "Competition in the Nuclear Power

Supply Industry," a report to the AEC and the U.S. Department of Justice (Washington, D.C.:U.S. Printing Office, December 1968); Irwin C. Bupp, Jr., "Priorities in the Nuclear Technology Program," Ph. D. thesis, Harvard University, Cambridge, Mass., 1971.

2. Craig Hosmer, telephone interview with IRRC, 1974.

3. Green, op. cit., p. 129.

4. Ibid., p. 141.

5. Charles R. Bardes, "Statement of Nuclear Energy Liability-Property Insurance Association to the Senate Committee on Public Utilities, Transit and Energy," Hearings on Provisions of the Nuclear Power Plants Initiative, Senate Committee on Public Utilities, Transit and Energy, California Legislature, January 27, 1976, p. 101.

6. Green, op. cit., p. 148.

7. John O'Leary, interview with IRRC, 1974.

8. L. Manning Muntzing, speech before the American Nuclear Society, 1972.

9. Eugene Cramer, interview with IRRC, 1974.

10. Victor Gilinsky, testimony before the Joint Committee on Atomic Energy, March 2, 1976, Vol. I, p. 306, "Investigation of Charges Relating to Nuclear Reactor Safety."

11. Interview with NRC officials, 1976.

12. O'Leary, op. cit.

13. Westinghouse comments to IRRC, March 16, 1976; U.S. Nuclear Regulatory Commission, Annual Report 1975. Also, Bernard Rusche, testimony before the Joint Committee on Atomic Energy, op. cit., p. 316.

14. Interview with IRRC, 1974.

15. Interview with NRC officials, 1976.

16. Telephone interview with NRC, December 1976.

17. Westinghouse remarks to IRRC, December 17, 1974, p. 29.

18. William H. Anders, testimony before the Joint Committee on Atomic Energy, op. cit., p. 151.

19. David Burnham, "AEC Files Show Effort to Conceal Safety Perils," New York *Times,* November 10, 1974, p. 1.

20. Anders, op. cit., p. 263.

21. David Burnham, "Atomic Energy Agency Completes Report Amid Charges it Stifles Criticism," New York *Times,* October 21, 1976.

22. Hal Willard, "NRC Acts on Safety Warnings," Washington *Post,* November 6, 1976, p. 1.

23. Ibid.

24. Richard A. McCormack, "Assessing Today's Nuclear Power Licensing Process," speech at the Atomic Industrial Forum, July 28–31, 1975.

25. GE comments to IRRC, January 10, 1975, p. 5.

26. Anders, op. cit., p. 151.

27. Romano Salvatori, "Nuclear Plant Safety," testimony before Michigan's House of Representatives, Lansing, October 24, 1974.

28. Marcus Rowden, letter to Ralph Nader, October 28, 1976.

29. George Stathakis, testimony before the Joint Committee on Atomic Energy, op. cit., p. 175.

30. McCormack, op. cit., p. 14.

31. Ralph Nader, testimony before the Joint Committee on Atomic Energy, January 28, 1974, pp. 1, 7, 8.

32. Green, op. cit., p. 152.

33. Anthony Roisman, interview with IRRC, 1974.

34. Daniel Ford, letter to IRRC, December 19, 1974.

35. Robert Gillette, "Nuclear Safety (I): The Roots of Dissent," *Science,* September 1, 1972, p. 771.

36. Burnham, "AEC Files Show Effort to Conceal Safety Perils," op. cit.

37. Jimmy Carter, "On Energy Reorganization," press release, September 21, 1976.

38. Anthony Roisman, Interview with IRRC, November 15, 1976.

39. See Harold Green, Hearings before the Subcommittee on Reorganization, Research and International Organizations, Senate Committee on Government Operations, March 12, 1976, pp. 172 ff.

40. Stephen Ebbin and Raphael Kasper, *Citizen Groups and the Nuclear Power Controversy* (Cambridge: MIT Press, 1974), pp. 210, 211.

41. Atomic Industrial Forum, "Causes of Nuclear Plant Delay," April 1974.

42. Anthony Roisman, interview with IRRC, 1974.

43. Common Cause, "Serving Two Master" (Washington, D.C., October 1976), pp. i. ii.

44. Ralph Nader, press release, October 25, 1976, available from Critical Mass, Washington, D.C.

45. Robert Pollard, testimony before the Joint Committee on Atomic Energy, February 18, 1976, Vol. I, pp. 97, 98.

46. Thomas O'Toole, "Atom Power Risks Said Disregarded," Washington *Post* October 21, 1976, p. A–7.

47. Burnham, "Atomic Agency Completes Report Amid Charges It Stifled Criticism," op. cit.

48. Ibid.

49. Dale G. Bridenbaugh, testimony before the Joint Committee on Atomic Energy, February 18, 1976, Vol. I, p. 25.

50. Dale G. Bridenbaugh, Richard B. Hubbard, and Gregory C. Minor, testimony before the Joint Committee on Atomic Energy, February 18, 1976, reprinted by the Union of Concerned Scientists, Cambridge, Mass., 1976, p. 43.

51. "NRC Declares Halt on Licenses for New A-Plants," Washington *Post,* August 14, 1976, p. A–16; "Nuclear Unit Moves to Resume New Plant Licenses," Wall Street *Journal,* October 14, 1976.

52. Carter, op. cit.

53. Thomas O'Toole, "Congress Considers Stripping Its Nuclear Panel of Power," Washington *Post,* November 7, 1976, p. A–13.

CHAPTER

5

THE GROWTH OF THE NUCLEAR INDUSTRY

Private industry—first the manufacturers of nuclear power plants and soon electric utilities as well—has responded favorably to the government's efforts to encourage civilian nuclear power development. Critics question the basic decision that industry has made to proceed with nuclear power development: "What I do not understand is why these large utility companies make such huge investments in nuclear power plants that may be unreliable, that they may not be able to use and [that] will not deliver the energy," Senator Abraham Ribicoff (D-Conn.) has commented.[1] Critics also question, as they do with government, the processes by which the decision was made and the provisions industry has made to tell the public of its decisions in a manner that would allow informed debate of the public-policy issues involved.

UTILITIES' COMMITMENT TO NUCLEAR POWER

Early Interest

Two factors, in addition to government incentives, encouraged utilities to invest in nuclear power plants. One, according to former Representative Craig Hosmer, was that utilities were looking for a competing source of fuel which would help keep down the price of coal and oil.[2] The other, according to MIT's Irwin Bupp, was that "the private utilities were extremely nervous about the evolving AEC power reactor development program. They were inclined to see it as potential camouflage for federal support of publicly owned utilities which in secret with the government would eventually 'squeeze' private utilities out of the power business with massive atomic TVAs."[3]

The 1960s

A report issued in 1968 by Arthur D. Little, Inc. noted that growing government experience with nuclear plants and rising fossil fuel prices were complemented by the willingness of GE and Westinghouse to supply and construct nuclear power plants on a fixed-price—or "turnkey"—basis. "The utilities needed the reassuring presence of known suppliers [GE and Westinghouse] who had demonstrated through past fossil projects technical resourcefulness in coping with unforeseeable difficulties, and reasonableness in sharing with the utilities the consequent burden of these difficulties,"[4] the study said. GE and Westinghouse, the report suggested, had built up large nuclear engineering divisions in support of government projects, and had a major stake in development of a civilian nuclear power industry.[5]

The result was an early, large commitment to nuclear technology. The report stated: "The fact is that utilities and manufacturers committed themselves to the installation of nearly 60,000 mw of generating capacity (representing an investment of over $10 billion) at a time when the only operating experience with nuclear plants was limited to installations which were, in some cases, only one-sixth the size of plants being ordered and having technical performance considerably less strenuous than the plants being ordered."[6]

The Decision-Making Process

The Arthur D. Little report described the process by which utilities decided to purchase nuclear power plants as follows: "In the typical utility process whereby a utility arrives at a choice of power plant supplier, the normal sequence of events is to proceed from a determination of the need for additional capacity, to the selection of a supplier, without an extensive intermediate phase of analyzing potential energy sources. The operating experience and the geography of the utility usually serve to pre-screen various energy sources, excluding some without the need for elaborate analysis."[7]

Many utilities argue that this is no longer the case, that they consider carefully the costs of using nuclear power and other available fuels. Eugene Cramer of Southern California Edison Company wrote to IRRC: "Whatever the situation may have been when A. D. Little wrote their report in 1968, USAEC nuclear licensing procedures require review both of alternate energy sources and techniques, as well as the need for this capacity."[8]

Officials of many utilities maintain that after analysis of the alternatives, which often involves sophisticated computer modeling of variables such as fuel and capital costs and plant reliability, nuclear power often appears to be the most economic for the utility and its customers. Duke Power Company, for example, told IRRC that a decision to build six nuclear units "will require us to market $841 million of additional new securities because of the larger investment in nuclear. However, the coal choice would have required us to collect from our customers $8 billion more in revenue over the lifetime of the six units than if nuclear. Good stewardship of the consumer's interest therefore suggested the nuclear choice if nuclear is acceptable from the environmental and safety points of view."[9]

Other utility officials admit that there are problems with nuclear plants, but they say the location of their systems and customers or their need for assured availability of power generation facilities gives them little choice but to go nuclear. They also say that they take any problems specific to nuclear power into account in their planning projections. Southern California Edison Vice President James Drake says that—because of the newness of nuclear technology—past performance does not provide an accurate indicator of the future. "Nuclear plants do represent a greater risk and have greater scheduling uncertainties," Drake comments, "and that economic risk is reflected in a larger allocation for contingencies in a nuclear construction budget."[10]

Status Today

The first civilian U.S. nuclear power plant began operation in 1957. By the end of 1970, 15 plants were operating, and by November 1976, 61 plants had operating licenses, representing 8.3 percent of U.S. generating capacity. Another 167 were under construction or had been ordered (see Figure 4).

FIGURE 4

Nuclear Power Plant Reactors in the United States

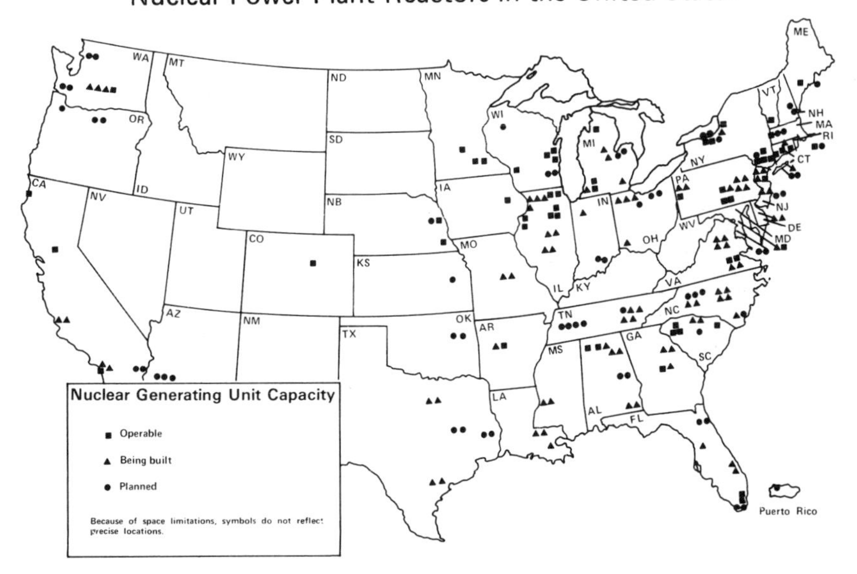

Source: U.S. Department of the Interior, *Energy Perspectives,* June 1976.

Moreover, the nuclear power plants being built now are larger in size.[11] Most plants in 1965 were less than 600 megawatts; most plants now being ordered are more than 1,000 megawatts.

Future Development

Some experts, including those within ERDA, project that by the year 2000 between 500 and 900 nuclear power plants with a total capacity of 500,000 to 9,000,000 megawatts will be in operation.[12]

Many utilities support the view that nuclear power is more economic than alternative forms of energy currently available. "At Commonwealth Edison, we believe nuclear power is the preferred method of large base-load generation today and out at least to the end of our planning horizons," R. L. Bolger wrote to IRRC in late 1975.[13] It is a belief shared by many persons in the industry.

Industry Dissenter

The major dissenter within the industry has been American Electric Power Company (AEP), owner of the largest privately owned group of electric utilities. AEP Vice President John Tillinghast told IRRC in 1974 that the company believed it would be some time before the operational and engineering problems with nuclear plants have been resolved. Until then, he said, the performance of nuclear plants will not be as good as has been projected. AEP has made its major investment in coal-fired plants. It has constructed two nuclear units—to gain some experience with nuclear technology, according to Tillinghast—but it has no plans to construct other nuclear units.[14]

Delays and Cancellations

Even those utilities that are bullish about nuclear power, however, have been reexamining their projected needs and commitments to build new power plants in light of tight money markets and reduced demand projections. Between March 1974 and June 1976, utilities removed from schedule 41 plants and suffered 161 delays of more than a year at plants planned or under construction. (Some of these delays have occurred at the same plants.) More than two-thirds of the delays and nearly two-thirds of the "removals from schedule" occurred at nuclear plants.[15]

Demand Pattern Shifts

Most utilities, in announcing delays or cutbacks, explain that their decisions are based in part on reassessments of growth in demand, reflecting conservation by consumers and industry, and on increased willingness of utilities to operate with less reserve capacity. "The load growth pattern or lack of it—since late 1973—has no precedent in utility history," reported *Electrical World* in September 1976. "The excursions we have experienced clearly show that traditional trending methods no longer apply, especially for the years in the immediate future."[16]

Financing Problems

In addition to uncertainties about demand, utilities have been plagued with uncertainties about their own ability to raise capital. Rate increases have not kept pace with rapidly rising capital costs. Utilities which had been able to meet 60 percent of their capital needs internally in the 1960s have been forced to borrow an increasing amount of their capital from external sources. Bankers Trust Company reports that in 1974 utilities reached the limit of the utility industry capital market and borrowed nearly 70 percent from external sources.[17] During the tight money market at this time, many utilities were unable to find sufficient capital to meet their projected plans and, as mentioned earlier, a number were forced to abandon or delay planned projects.

The capital market has eased recently, and public service commissions have granted rate increases to many utilities. Nevertheless, Bankers Trust estimates that utilities will depend on external financing to meet 65 percent of their investment needs in the future. The bank comments that "at a factor this high, the industry will remain very sensitive to interest rates, regulatory lags and delays, and increasing costs of construction."[18]

Many utilities faced with capital shortages are more inclined to cut back on plans to construct new nuclear power plants. Because nuclear plants are more expensive to build, cancellation of a nuclear plant allows a greater savings. Also, there has been a long time lag—ten years—between the decision to build a nuclear plant and start-up of operations. Thus utilities are inclined to postpone construction of nuclear plants for at least two or three years, during which time they can reassess changing trends in demand. If it appears that growth in demand is resuming at close to historic rates, they can meet it by constructing coal-fired plants, which have a shorter (five- to six-year) lag time. "The uncertainties make you pause and say, 'gee, let me wait another year,' " according to Clyde A. Lilly, former president of The Southern Company Services Inc. "And if you do that a couple of times, then you don't have time to put in a nuclear unit."[19]

In sum, the capital shortage, together with uncertainties about demand, has led to a partial moratorium on nuclear power development. Industry representatives believe this may lead to serious problems in the mid-1980s, but intervenors and critics of nuclear power have viewed it as an opportunity to brake the momentum of nuclear power development.

THE MANUFACTURERS' ROLE

As noted earlier, the existence of GE and Westinghouse as suppliers of nuclear power was an important factor in utilities' decisions to buy nuclear power plants. The number of manufacturers is still small (GE, Westinghouse, Combustion Engineering, Babcock & Wilcox, and General Atomic), and the influence the manufacturers wield is still very strong. Because of their expertise, they have operated at an advantage as sellers, and utilities have turned to them for information on plant operation and for projections of power plant reliability. And because there are so few of them, they have been able pretty much to establish the terms of warranties.

Projections

Many utility officials admit that they are largely dependent on manufacturers for information about nuclear plant operations. They generally agree that manufacturers' projections of the expected capacity factor of nuclear plants, costs of fuel and operation, and maintenance costs have been optimistic. "It's painfully true," the vice chairman of one large utility told IRRC, "that manufacturers' projections have been overstated. But where else do you go?" Most utilities—with a few notable exceptions like Commonwealth Edison—have limited experience with nuclear power. "Of 9,000 presidents, chief executives and chief engineers in utilities," Milton Levenson of the Electric Power Research Institute noted in 1974, "less than 1 percent have a nuclear background."[20]

Warranties

A major problem for utilities that arises out of this "seller's market" is the problem of obtaining warranties from the manufacturers which come close to covering potential losses to the utilities. W. H. Arnold of Westinghouse explained his company's position on warranties to IRRC. He said the basic warranty runs for one year from the date of "plant acceptance," provided the equipment is installed in accordance with Westinghouse specifications. Under the warranty, he said, Westinghouse will fix or replace any defective equipment it supplied. It will not pay for damages to other equipment or for consequential damages such as the cost of buying power from other utilities to replace power that cannot be generated as a result of equipment failure.

According to Arnold, Westinghouse is "not very flexible" about negotiating the basic terms of the warranty. The nuclear reactor, he said, represents only about 10 percent of the value of the plant. For Westinghouse to assume enormous contingent liabilities when its involvement is so limited would not be prudent, he asserted. However, he noted, on the question of who bears the risk of loss due to changes in regulatory requirements, the company is willing to assume some responsibility.[21]

The question of warranties is further confused by changing safety or environmental standards for operating nuclear plants. Some vendors say they should not be expected to provide warranties for a plant's ability to meet licensing requirements beyond the plant start-up date. "The vendors can't do it [provide extended warranties] as responsible and professional businessmen," writes General Atomic's former president Richard McCormack, "because they do not, cannot and should not set safety criteria and standards. The fundamental issue is beyond their control and therefore has no place in commercial terms and conditions." McCormack argues that safety costs are established by government, through the setting of standards, and "for this reason I believe that there is real merit in providing NRC with the budget required to pay for justificable ratcheting, so that the agency can pursue its best safety judgment without concern for industry pressure."[22]

Other industry officials told IRRC that the manufacturers cannot be expected to increase warranty protection unless they have a great deal of control over how the power plant is built and operated. The utilities do not want to relinquish control, particularly with respect to plant operations. Thus, they must accept more limited warranty protection.

Utility-Vendor Conflicts

The relationship between utilities and vendors has been affected by claims vendors have made about the performance of the plants, "sweeteners" in terms of contracts to supply or reprocess fuel that have been included in the sales agreements, and limits on the warranties covering plant operation. Changes in the industry—the sudden rise in the price of uranium, the failure of a new reprocessing technology, or the instituting of new standards—have brought increased and unexpected costs to both utilities and manufacturers. There have been disputes and, in several cases, suits, over who should shoulder the new costs. In 1975, as reported earlier, Westinghouse withdrew from uranium supply contracts with 27 utilities when the price of uranium jumped. The 27 utilities have sued Westinghouse; Westinghouse, in turn, has initiated action against 29 firms which it accuses of illegally raising and fixing prices.[23] Several utilities have initiated claims against General Electric for failure to fulfill its contracts for reprocessing and storage of plant wastes after it abandoned its efforts to get its Morris, Ill., reprocessing plant into operation. Several utilities have sued vendor manufacturers, architect-engineers, and others because nuclear plants have failed to operate at the promised capacity levels.[24]

Just as there is debate over whether manufacturers or utilities should be held responsible for unforeseen costs of generating power, there is also conflict over whether shareholders or consumers should make up the difference between planned and actual costs. Some public service commissions have allowed the utilities to pass the costs on directly to consumers in the form of a fuel adjustment clause.[25] But others have begun to argue that utilities' stockholders should cover the costs, either by reclaiming costs from the manufacturer or by absorbing them. Two of three Wisconsin public service commissioners asserted in August 1976 that Wisconsin Electric Power had erred in not writing a penalty clause into its fuel reprocessing contract with GE. When GE was unable to fulfill the contract, the commissioners said, the utility and its stockholders either had to get restitution from GE or absorb the costs.[26] As a result of these unexpected costs, a number of commissions have begun to question utility license applications more closely and to demand a more detailed analysis and justification for the construction of nuclear plants.[27]

INFORMING AND PERSUADING THE PUBLIC

In general, the efforts made by industry to inform the public of its nuclear power program have concentrated on trying to persuade members of the public to accept those programs. Until the mid-1970s, there appeared to be little need for a major industry effort to discuss the pros and cons of nuclear power. Few questioned statements by reactor vendors and government officials that nuclear power would be the predominant source of electrical energy in the future. Most utilities accepted vendors' claims and left response to questions about nuclear power to government information officers, vendors, or organizations financed by industry such as the Atomic Industrial Forum. The situation has changed dramatically in the last two years, as public debate over nuclear power has escalated.

The Situation in 1974

A fairly accurate assessment of industry's information programs as late as 1974 can be made on the basis of responses that IRRC received to a questionnaire it sent to 82 companies involved in all phases of the industry. The questionnaire asked for data on companies' experience with nuclear power, cost analyses, and copies of statements relevant to the elements of the controversy over nuclear power development.[28] Virtually all of the materials IRRC received that had been prepared for distribution to the general public were public relations pieces emphasizing the merits of nuclear power. The only major issue of public concern that was discussed meaningfully in these materials by several companies was power plant safety. The one exception to this trend was General Atomic, which had prepared several thoughtful papers directed at a general audience, including a detailed discussion of the relative responsibilities of industry and government for safeguarding special nuclear materials.

The materials prepared for trade audiences similarly failed to address what W. H. Arnold of Westinghouse has called the "societal issues" raised by nuclear power development. Most appeared to be efforts to convert the converted. A few publications, notably Westinghouse's *Nuclear Energy Digest,* contained a good deal of general and technical data and clearly had been prepared for distribution within the industry. None of the published materials that IRRC received included any extensive comparison of costs of nuclear and coal-fired power plants, or any examination of alternatives to nuclear power development.

Utilities, for the most part, did not address policy issues. Duquesne Light and Power Company, operator of the first commercial nuclear power plant in the United States, wrote that it "has not made any public policy statements regarding Price-Anderson, standardization, siting, safeguards or waste disposal." The company did answer questions relating directly to reactor operations—on operational safety, licensing, and the development of the breeder reactor. Consolidated Edison Company, Philadelphia Electric Company, and the Sacramento Municipal Utility District suggested that IRRC direct its inquiries to the Atomic Industrial Forum and the Edison Electric Institute. Cincinnati Gas and Electric Company said it had made "no formal public statements regarding any of the other [policy] categories listed, but in news interviews and similar discussions, we have expressed our conviction that nuclear power can and must be developed as a safe and reliable energy source." Illinois Power Company stated simply that it had not made "public statements on the various subjects in the areas listed in your questionnaire."

A few utilities did indicate that they followed policies of reacting to issues raised by others. For example, Daniel Green of Niagara Mohawk Power Company described his company's policy as "quite aggressive." He wrote: "We have participated in a host of television programs related to nuclear power. In addition, we fought vigorously a nuclear moratorium initiative proposed by a statewide local government organization, which was subsequently defeated. We have tried not to let any unfavorable editorial or article go unanswered in the print media, and we have spoken frequently to a host of civic, fraternal and social organizations."

Recent Changes

In the last two years, however, there has been a dramatic change in information efforts by members of the nuclear industry. Some utilities, such as Commonwealth Edison, and reactor vendors have increased the amount of information they provide to the public. They have designed materials that address specifically concerns about safety and reliability of nuclear plants, and they have discussed publicly problems with reprocessing and waste disposal. The thrust of industry's public information campaign continues to be persuading the public of the benefits of nuclear power, but the quantity of the material has increased substantially.

The growth of industry materials is a direct response to the success of nuclear opponents in putting issues relating to nuclear development before the voters or their representatives. In 1976, initiatives to slow or curtail nuclear development appeared on the ballots in seven states; in the near future they are likely to be presented to voters in at least nine other states. Antinuclear bills have been considered in more than 20 state legislatures, and three have become law in California. At the same time, critics have increased their expertise and have expanded their attack on nuclear power. Not only are they continuing to challenge industry assertions about the acceptability of nuclear power, but they are attacking it at its very heart: its claims of economic viability. Armed with new manuals prepared by environmental and public interest groups, they are questioning utilities' statistics on the potential savings for consumers, the capacities of nuclear plants and the future levels of demand for power.[29]

Industry has responded to the critics' challenge in several ways. In a number of states, representatives of the nuclear industry, business organizations, scientific associations, and labor have joined together to support groups, such as California's Citizens for Jobs and Energy, which recommend development of nuclear power as one aspect of a comprehensive energy plan. A halt to nuclear power, they argue, will have a serious economic impact, costing thousands of workers' jobs.[30] Such groups provided the backbone of industry's successful efforts to defeat the antinuclear initiatives. And utilities as well as vendor manufacturers made substantial contributions to their campaigns. In Ohio, utilities had spent nearly a million dollars in the 1976 elections, for example, in support of efforts to defeat antinuclear initiatives.[31]

Westinghouse and other companies have come to view public information as critical to the future of nuclear power. "We view the major problem which must be overcome to be educating the public to the point at which they can rationally weigh the arguments of both opponents and proponents of nuclear power," a Westinghouse representative wrote IRRC in March 1976. "The nuclear industry firmly believes that once the public understands nuclear power and the manner in which nuclear plants are designed, built and operated, the great majority of the issues now in public controversy will be laid to rest. This is the reason why the industry is now putting greater emphasis on public information systems."[32]

In addition to building up their own information systems, utilities and industry, through the Atomic Industrial Forum and other representative groups, have increased the tempo of lobbying efforts on behalf of nuclear development. In 1975 the Atomic Industrial Forum moved to Washington and helped to establish a lobbying organization called the American Nuclear Energy Council. At the same time, industry has been called upon to support other pronuclear groups such as Citizens for Jobs and Energy and Americans for Energy Independence.

THE CONTROVERSY OVER INDUSTRY'S ROLE

Critics of the nuclear industry's role in nuclear power development raise three principal issues: (1) that the manufacturers have misled the utilities in their projections, and concurrently that the utilities have failed to challenge the manufacturers' projections; (2) that the utilities have jumped into nuclear power too fast, without adequate consideration of the alternatives or the consequences; and (3) that both manufacturers and utilities have failed to inform the public adequately of their programs and the problems involved.

Manufacturers' Assessments

Manufacturers in general agree with the criticism that some of their assessments have been too optimistic and that there have been problems with many plants, but they do not agree with assertions that they have deliberately misled the utilities. The manufacturers assert that most of the problems have been unforeseen and that many have arisen as a consequence of either regulatory decisions or inadequate performance by others in constructing or operating the plants. Some observers point to the high losses the manufacturers have sustained—particularly on turnkey projects—as added evidence that the problems with nuclear plants were as much a surprise to the manufacturers as to the utilities. One former industry marketing executive has estimated that GE and Westinghouse together suffered losses of between $800 million and $1 billion on 18 plants built on a turnkey basis.

In November 1976 the *Wall Street Journal* reported that GE intended to revise its nuclear marketing strategy in order to limit the company's liability and to improve plant reliability. According to the *Journal,* GE is considering moving to cost-plus contracts on sales and a system of shared responsibility for nuclear plant reliability among companies, including GE, which supply equipment for the plant. The result, some say, would be better—if somewhat more expensive—nuclear plants. "When you're losing money or getting an inadequate return on a sale, there's always the temptation to avoid doing those little extra things that often mean a much more reliable plant."[33]

The Nuclear Commitment

The utilities' desire to go nuclear, according to some critics, is attributable at least in part to financial motives. Utility rates are computed by public service commissions to allow for a certain rate of return on capital investment. Critics say that because nuclear plants are relatively capital-intensive, this method of computing rates makes it financially attractive for a utility to buy a nuclear power plant, rather than a coal-fired plant that will cost less and have somewhat higher operating costs. As a result, utilities have been too willing to accept manufacturers' projections of the efficiency of nuclear power and have committed themselves prematurely to even larger nuclear plants.

A number of observers agree that the industry has gone too far too fast. Harold Green criticizes what he describes as a "leapfrogging technical experience,"[34] and James Conner, then director of planning and analysis for the AEC, wrote in 1973 that "in retrospect it is clear that the initial cost and time schedules for nuclear power plants were wildly optimistic, given the industry's lack of experience with the technology. . . . Since the first units were ordered, average plant size has increased by over 90 percent. Utility schedules have had to be lengthened by over 40 percent to accommodate the delays attributable to increasing size and evolving technology." The new technology posed serious problems for utilities' managers. "Most unsettling of all," Conner relates, "utility executives have discovered that they are responsible for these troublesome new plants. By law, the AEC licenses the utility itself—not the manufacturer—to build and assure safe operation of a nuclear plant."[35]

Some observers are more critical. Irwin Bupp of MIT describes the prevalence of a "stainless steel mentality." Utility officials, he says, need to feel that they are in the forefront of the industry; they believe that coal technology represents the 19th century and that fission technology represents the science of the future.[36]

Others are more charitable. Several point to uncertainties about the prices and availability of fossil fuels as having played a major role in utilities' thinking. "We know we have problems with supply from the Arabs or the United Mine Workers," one official told IRRC. He and others are more hopeful that problems with nuclear power can be solved.

Informing the Public

Critics assert that the nuclear industry has shared with government a failure to keep the public adequately informed on the problems of nuclear power. "There has been a tacit conspiracy on the part of the atomic energy establishment—industry and government—for the last 20 years to hide from the public view the risks inherent in nuclear power," Harold Green has testified. "I do not use the phrase 'conspiracy' in an invidious sense. The fact of the matter is that the establishment fears that the public discussion of the risks will unduly alarm the public and slow the introduction of nuclear power which the establishment honestly believes is acceptably safe and in the public interest."[37]

What critics see as a failure by industry to disclose information on nuclear power may be seen by industry—at least in part—as a proper withholding of proprietary information. Critics, however, argue that industry has hidden behind claims of confidentiality in order to avoid discussing its problems openly.

The information gap may also reflect an unwillingness on the part of industry or government officials to take critics seriously. Some supporters of nuclear power agree with critics that industry has simply avoided confronting public concerns directly. Six proponents of nuclear power issued a paper in 1975 reprimanding both critics and industry for not providing the public with "the objective and unemotional facts that the public rightly deserves." Some members of industry, they said, "feel the public deserves no better than a childish argument and an elitist 'we know better' attitude."[38] Dale Bridenbaugh, one of the three engineers who resigned from GE, said in explaining his departure from GE that "one of the things that prompted us to speak out is the fact that there has been a lot of public concern expressed about nuclear power but the standard response by the industry spokesmen has been, 'well, that is uninformed people worrying about things, and there are no technical people in the industry that are concerned.' "[39]

Perhaps the overriding complaint of critics is that in its discussion of nuclear power, the industry has focused on selling its benefits to the public, rather than on assisting the public to reach an understanding of its implications. The desire to sell, some say, has encouraged industry to mislead. As mentioned earlier, in 1975 the Council on Economic Priorities challenged Consolidated Edison's claims that the company's nuclear plants had saved consumers $90 million. The claims, according to the council, failed to take into consideration capital and operation costs.[40] Although the company maintained that it was clear that its statement only referred to savings in fuel costs, New York Public Service Commissioner Kahn sided with the council. In a letter to Con Ed's chairman, he said he could not see how the company's claims "could fail in some degree to mislead the public" about the comparative costs of nuclear and fossil fuel costs.[41]

Concern with industry's efforts to sell nuclear power is compounded by its ability to spend large sums on its own behalf. Critics were particularly bitter about the size of corporate contributions in opposition to the antinuclear initiatives on the ballots in seven states across the country in 1976. In California, Pacific Gas & Electric spent $500,000 and other power companies contributed more than $300,000 in a successful effort to defeat a proposition which would have slowed nuclear growth.[42] In Arizona, groups favoring nuclear power—including labor and professional as well as industry donors—spent $1.1 million compared with $17,000 for those opposed to nuclear power. In Oregon the spending ratio was 4 to 1, and in Montana nuclear proponents spent $120,000 to the $600 spent by nuclear critics.[43] The elections, critics stated, were less a debate of nuclear power than a demonstration of the power of the nuclear industry's advertising budgets.

Supporters of nuclear power agree that they have not been able to provide information sufficient to convince the general public of the benefits of nuclear power, but they deny vigorously all charges that they have withheld important facts or misled the public. They argue that critics of nuclear power have been irresponsible in their statements, using sensationalism to attract public attention, and making thoughtful public discussion of the issues very difficult.

Many persons in industry say that the public mood and the nature of the news media are not conducive to a fair hearing of the industry's position. H. D. Hexamer, manager of GE's department of communications and nuclear power information, wrote to IRRC in 1975:

> It's common knowledge . . . that during the past 25 years, popular respect for science and its corollary congressional support reached its apogee shortly after World War II, and has continually declined since then. Many of the critics . . . have been demonstratively adept at sensing this aspect of the public mood and have chosen to place the blame for pollution, energy shortages, and other national ills on science and technology. Of course they have reinforced their efforts with attacks on the Atomic Energy Commission, the utility industry, Congress, and a sensationalist discussion of the various so-called safety problems of radiation, reactor emergency cooling, and so forth. . . . [As a result,] it's industry's position the evidence is very strong that the determined citizen who wishes to be heard, particularly if he opposes the government, a utility, or the establishment in any form, gets attention, magnified by a receptive press, far out of proportion to his single vote. The industry believes the need is not so much for just the opportunity to speak but rather for an informed and active public willing to responsibly and thoughtfully express its opinion.[44]

It is also difficult to generate positive news coverage, industry spokesmen say. "It is much more likely that a newspaper will be interested in running a headline describing cracks in a minor steam pipe at a nuclear plant than reporting that the plant is operating at full capacity and saving consumers' dollars," a Commonwealth Edison official told IRRC.[45]

Another problem, according to industry officials, is that utilities often are ill-equipped to handle public relations questions. A representative of Commonwealth Edison told IRRC: "Our job has been to run a business and we have done so with the belief that the Edison Electric Institute or the Atomic Industrial Forum should fight the battles. Maybe we should set up a department to fight battles," he said. "But we have to decide where to spend our shareholders' money—on public relations or on day-to-day business."[46]

Similarly, Eugene Cramer of Southern California Edison wrote IRRC that "since public utilities commissions in general have proscribed in recent years any extensive mass communications, it is doubtful if utilities have been able to afford the luxury of more than an occasional weak defense against a high-bill complaint. It should also be recognized that a utility would logically tend to provide excellent information readily available [from a trade or professional source], rather than reinventing the wheel."[47] "The tendency still," according to Paul Dragoumis of Potomac Electric Power Company, "is to leave public discussion of nuclear policy issues to the Atomic Industrial Forum or other industry organization which is not really constituted to deal with policy analysis or formulation." Another constraint on utility involvement on nuclear policy issues, in addition to the limitation placed by public regulatory commissions, is concern about a possible challenge by nuclear opponents seeking equal air time. Ultimately, Dragoumis commented, the utility industry's response to its financial problems was to cut its budget for communications as the financial crisis deepened.[48]

Finally, as concerns technical information, industry spokesmen argue that they must have the right to withhold some information which they consider to be proprietary. Spokesmen say industry is willing to make available to the public all necessary information on plants and operation. But they agree with Westinghouse's Arnold that "unrestricted disclosure of proprietary information would harm significantly the interests of the commission and the public. Unrestricted disclosure could discourage initiation of research and development by private parties, limit the knowledge of the existence of such information, impair the commission's independent review process, and endanger the position of the U.S. as the world leader in nuclear power reactor technology and adversely influence our growing contribution to the U.S. balance of payments."[49]

GE and Westinghouse have told IRRC that they have made proprietary information available to the NRC, to the Joint Committee, or even to intervenors "under appropriate protective agreement."[50]

The last two years have seen some change in the quantity and quality of information available from industry representatives. The initiative campaigns have spurred the quantity of material produced by companies interested in nuclear development. Many of the publications produced in response to the initiatives reflect a traditional public relations approach to the issues, but as members of the public have become more sophisticated in their concerns about nuclear power, industry representatives have become more sophisticated in their treatment of the issues. Whether the increased sophistication in the discussion of issues has resulted in a more candid exchange of information between nuclear proponents and their critics, however, remains doubtful.

NOTES

1. Abraham Ribicoff, Hearings before the Subcommittee on Reorganization, Research and International Organizations, of the Committee on Government Operations, U.S. Senate, December 5, 1973, p. 218.
2. Craig Hosmer, interview with IRRC, 1974.
3. Irwin C. Bupp, Jr., "Priorities in the Nuclear Technology Program," Ph.D. dissertation, Harvard University, Cambridge, Mass., 1971.
4. Arthur D. Little, Inc., "Competition in the Nuclear Power Supply Industry," report to the U.S. Atomic Energy Commission and the U.S. Department of Justice, December 1968, p. 361.
5. Ibid.
6. Ibid., p. 124.
7. Ibid., p. 358.
8. Eugene Cramer, letter to IRRC, January 3, 1975.
9. Duke Power, letter to IRRC, July 1974.
10. James Drake, interview with IRRC, 1974.
11. Atomic Industrial Forum, Press Info, Number 70, November 1976.
12. U.S. Nuclear Regulatory Commission, "Final Generic Environmental Statement on the Use of Recycled Plutonium in Mixed Oxide Fuel in Light Water Cooled Reactors, Executive Summary," (Springfield, Va.: National Technical Information Service, August 1976), p. S–12.
13. R. L. Bolger, letter to IRRC, December 23, 1975.
14. John Tillinghast, interview with IRRC, 1974; AEP Public Relations Official, phone conversation, November 1976.
15. Edison Electric Institute figures, phone conversation, November 1976.
16. "Forecast," *Electrical World,* September 15, 1976, p. 54.
17. Bankers Trust Company, Energy Group, "Capital Resources for Energy Through the Year 1990," (New York, 1976), p. 23.
18. Ibid.
19. Quoted in "Why Atomic Power Dims Today," *Business Week,* November 17, 1975, p. 98.
20. Milton Levenson, interview with IRRC, 1974.
21. W. H. Arnold, interview with IRRC, 1974.
22. Richard H. McCormack, "Assessing Today's Nuclear Power Licensing Process," speech to the Atomic Industrial Forum, Seattle, July 28–31, 1975, p. 14.
23. Eric Morgenthaler, "Westinghouse's Woes Over Uranium Supplies Affect Entire Industry," *Wall Street Journal,* September 30, 1976; also, Byron E. Calme, "Westinghouse Charges 29 Firms in Uranium Suit," *Wall Street Journal,* October 18, 1976.
24. "Nebraska Public Power Files Lawsuit Against Everybody," *Not Man Apart,* November 1975; "Consumers Power Suit Against Bechtel and Three Others," *Not Man Apart,* mid-November 1974; "Vermont Yankee Nuclear Co. Settles," *Public Power Weekly,* January 30, 1976.
25. Milwaukee *Journal,* August 12, 1975, p. 1.
26. *Electrical World,* August 30, 1976, p. 3.
27. Newark *State Ledger,* February 6, 1975, p. 1; *Nucleonics Week,* October 30, 1975, p. 2.
28. Responses to IRRC questionnaire, 1974.
29. See Ron Lanoue, *Nuclear Plants, The More They Build, The More You Pay* (Washington, D.C.: Center for Study of Responsive Law, 1976); Richard Morgan, *Phantom Taxes in Your*

Electric Bill, and *A Citizen's Guide to The Fuel Adjustment Clause,* Utility Project of the Environmental Action Foundation, September 1975.

30. Michael Peevey, "Economic Impact of the Nuclear Power Initiative," hearings before the California Senate Committee on Public Utilities, Transit and Energy, Los Angeles, February 20, 1976, p. 15.

31. Herb Epstein, *Critical Mass,* Washington D.C., February 11, 1977.

32. Westinghouse, letter to IRRC, March 17, 1976, p. 14.

33. Tim Metz, "GE Studies Nuclear Marketing Changes to Limit Liability, Raise Profit Potential," *Wall Street Journal,* November 18, 1976, p. 2.

34. Harold P. Green, Hearings before the Subcommittee on Reorganization, Research, and International Organizations of the Committee on Government Operations," U.S. Senate, March 12, 1974.

35. James E. Conner, "Prospects for Nuclear Power," reprint from "The National Energy Problem," *Proceedings of the Academy of Political Science* 31 (December 1973): 66, 67.

36. Telephone conversation with IRRC, 1974.

37. Green, op. cit., p. 229.

38. Ian A. Forbes et al., "The Nuclear Debate: A Call to Reason," reprinted by California Council for Environmental and Economic Balance (San Francisco, September 1975), p. 1.

39. Dale G. Bridenbaugh, Hearings before the Joint Committee on Atomic Energy, February 18, 1976, Washington, D.C., p. 74.

40. Charles Komanoff, "Responding to Con Edison: An Analysis of the 1974 Costs of Indian Point and Alternatives," Council on Economic Priority, 1975.

41. Alfred Kahn, letter to Charles Luce, February 27, 1976.

42. "Money From Dominoes," *The Power Line* (Environmental Action Foundation): 2, no. 1 (June/July 1976): 3.

43. Figures from the Utility Project, Environmental Action Foundation (Washington, D.C., November 1976).

44. Hugh Hexamer, letter to IRRC, January 10, 1975.

45. Telephone interview with IRRC, 1974.

46. Ibid.

47. Cramer, op. cit.

48. Telephone interview, November 30, 1976.

49. Interview with IRRC, 1974.

50. Ibid.; see also GE's testimony, Hearings before the Joint Committee on Atomic Energy, February 24, 1976, pp. 170 ff.

CHAPTER

6

THE STATUS OF NUCLEAR TECHNOLOGY

Virtually all of the 228 nuclear power plants that are operating, under construction, or on order in the United States use "light-water" reactors to generate nuclear power. Eventually, a few may use "high-temperature, gas-cooled" reactors, but these reactors have been withdrawn—at least temporarily—from the commercial marketplace, and only one is now operating on a demonstration basis.[1] For the fairly distant future, government and industry research and development efforts are concentrating on the "breeder" reactor, which is expected to "breed" nuclear fuel and thus provide a virtually endless supply of energy.

Because the safety issues raised by nuclear power opponents relate in large part to the ways nuclear power is generated, this chapter outlines the status of nuclear technology. Subsequent chapters will discuss the potential consequences of use of nuclear technology and the steps that are being taken or could be taken to prevent adverse consequences.[2]

THE LIGHT-WATER REACTORS

There are two kinds of light-water reactors: boiling-water reactors (which are made by General Electric), and pressurized-water reactors (which are manufactured by Babcock & Wilcox, Combustion Engineering, and Westinghouse). Light-water reactors run on energy created by the fission, or splitting, of uranium atoms, which throw off two or three neutrons and release heat in the process. The neutrons in turn run into other atoms, dislodging more neutrons, and, if properly controlled, setting up a chain reaction and providing a steady source of heat.

The light-water reactors now in operation use "enriched" uranium, containing about 3 percent of the isotope U-235, as their fuel. Natural uranium contains less than 1 percent U-235; almost all the remainder is U-238. The proportion of U-235 is increased through a complex gaseous diffusion process at special ERDA facilities. (Weapons-grade uranium is enriched to more than 90 percent U-235.)

For use as a reactor fuel, the slightly enriched uranium is oxidized and formed into fuel pellets, which are placed into 12-foot rods clad in zirconium. The rods are assembled in bundles and placed vertically in the reactor core. The reactor core is surrounded by a heavy steel pressure vessel, and the pressure vessel and supporting cooling systems are placed inside a large reinforced-concrete containment dome. The chain reaction is controlled by rods —usually made of boron—which are partially raised from the reactor core and by boric acid which circulates in water around the fuel bundles. Boron is a neutron-absorbing element; when the rods are dropped into the fuel core, the boron absorbs neutrons being thrown off by the splitting atoms and controls the chain reaction. Removal of the control rods from the core starts up the reactor; thrusting them back into the core terminates, or "scrams," the reactor's operation.

The reactor operates until the fuel is "spent," when it is no longer able to sustain a chain reaction economically because it has used up a major portion of its fissile material and has accumulated too many neutron-absorbing byproducts. The spent fuel is comprised of a variety of items, including plutonium produced through the fissioning of U-235, amounts of unused U-235 and smaller quantities of U-238, and radioactive wastes. The plutonium, which itself is highly radioactive, and uranium may later be extracted from the unusable wastes to be recycled and made into fuel for light-water or breeder reactors.

The Boiling-Water Reactor

In a boiling-water reactor, the heat from the fission process—contained in the reactor core—boils water that is passing through the core and thus creates steam (See Figure 5). The steam drives the turbines that generate the electricity. After the steam passes through the turbines, it is condensed, and the water formed is returned to the reactor core. There, the water is boiled again, cooling the reactor core in the process, and so the cycle continues.

The Pressurized-Water Reactor

A pressurized-water reactor operates in much the same fashion, except that the water is kept under pressure, which prevents it from becoming steam, and the radioactive coolant is isolated from the steam turbine by a heat exchanger (see Figure 8). The heated water flows through the heat exchanger,

FIGURE 5

Boiling-Water Reactor Power Plant

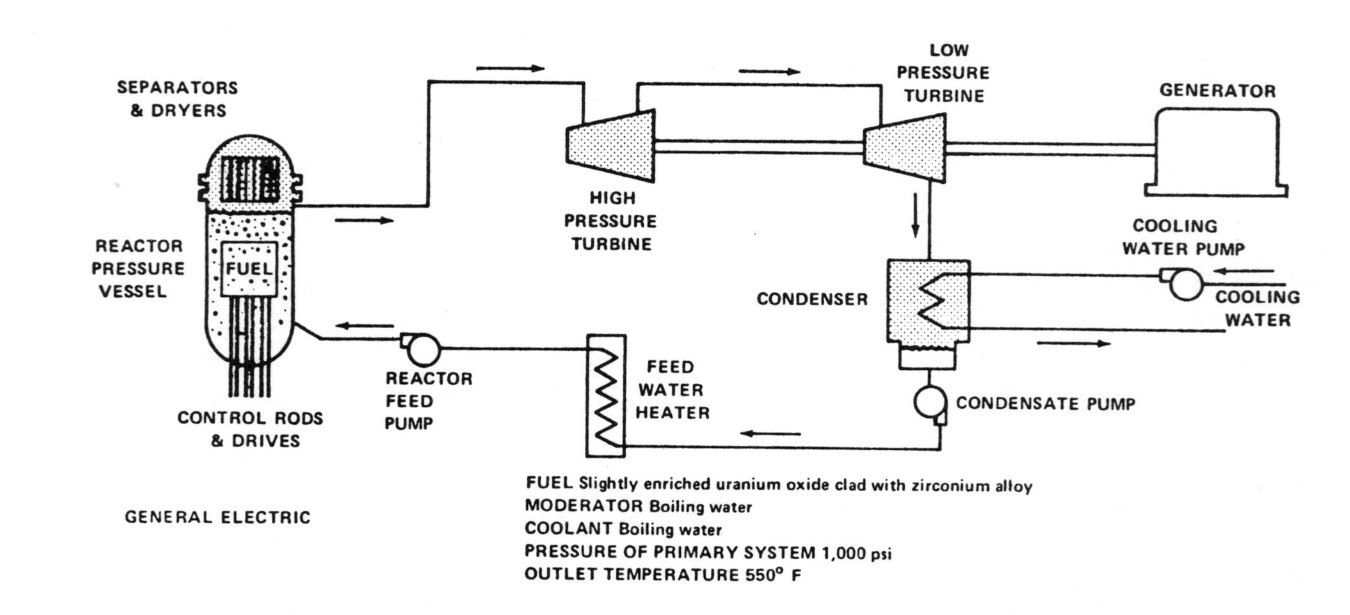

Source: Atomic Energy Commission, *"The Nuclear Industry 1974,"* WASH 1174–74.

FIGURE 6

Pressurized-Water Reactor Power Plant

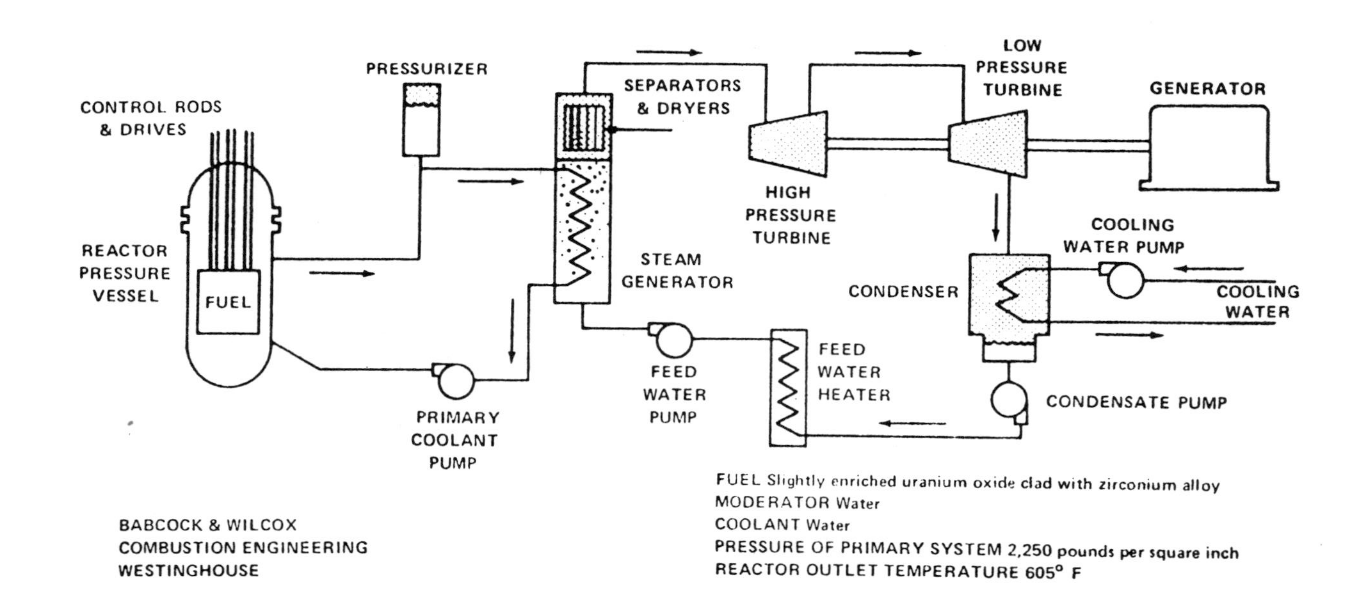

Source: Atomic Energy Commission, *"The Nuclear Industry 1974,"* WASH 1174–74.

where it is cooled, and water in a secondary system is boiled to create steam to drive the turbines.

Although most diagrams of light-water reactors make them appear quite simple in design, and nuclear power plants often are said to operate in essentially the same manner as fossil fuel plants, in fact the design of nuclear power plants is very complex. A maze of pipes is needed for the primary-water, cooling-water, and emergency-core-cooling systems. In discussions with IRC in 1974, an AEC official emphasized that it was inappropriate to compare nuclear and fossil fuel plants. Nuclear plants, he said, are much harder to build and require special cooling systems and containments not required by fossil plants.

THE HIGH-TEMPERATURE, GAS-COOLED REACTOR

The Energy Research and Development Administration has under way a program to develop high-temperature, gas-cooled reactors (HTGRs) and to test them for commercial marketability. The reactors, originally designed by General Atomic Corporation, were removed from the commercial market in 1975 when General Atomic encountered financial and technical difficulties. One 330-megawatt plant is being used as a demonstration plant, and orders for eight others have been dropped. ERDA hopes that "by the mid-1980s the technology base necessary for construction and operation of large commercial HTGRs" can be completed and that initial commercial operations can begin in the 1990s.[3]

The HTGR differs in several important respects from the light-water reactor:

Fuel: Highly enriched uranium 235 mixed with thorium is formed into cylindrical rods about one-half inch in diameter by two inches long and inserted into graphite blocks. Thorium is more naturally abundant than uranium.

Coolant: Pressurized helium is circulated through cylindrical passages in the graphite of the reactor core. The helium carries the heat from the reactor through a heat exchanger which in turn generates the steam to drive the turbines.

Fission products: During the fission process, uranium 233 is produced from the thorium, which captures neutrons from the uranium 235. Spent fuel assemblies are reprocessed to extract the U-233, which is then recycled for use as fuel.

The arguments that have arisen over certain technological aspects of the HTGR can be summarized as follows:

Arguments for the HTGR: Proponents of the HTGR say the following characteristics make it superior to the light-water reactor: (1) It is more efficient in converting energy to power—40 percent compared with 32.5 percent for the light-water reactor—and uses 40 to 50 percent less uranium than a light-water reactor of equivalent size. (2) Its higher thermal efficiency means less thermal pollution, which could save up to 9 million gallons of water a day. (3) Use of graphite as a moderator increases safety and eliminates the need for an emergency cooling system in the event of accidental rupture of the coolant line. (4) It produces less plutonium and thus reduces plutonium safeguards and waste disposal hazards.

Arguments against the HTGR: Those who remain skeptical about the HTGR say: (1) While the use of graphite eliminates the need for an emergency cooling system, the HTGR still requires a second means of forced circulation of the coolant in emergency situations. (2) Although the HTGR produces no plutonium, it uses highly enriched uranium 235 and produces uranium 233, both of which are weapons-grade material. Like the light-water reactor, it produces long-lived fission products in its wastes. (3) The HTGR's efficiency should not be compared with the light-water reactor but with the breeder reactors. (Proponents of the breeder believe it is economically more attractive than the HTGR, although neither system has yet had any commercial operating experience.)

THE BREEDER REACTOR

Some of the most controversial issues in the nuclear power debate involve the liquid-metal fast-breeder reactor (LMFBR). The LMFBR produces, or "breeds," more fuel than it consumes. Plutonium, initially produced in light-water reactors, is used together with uranium 238 in the fuel assembly. The fission process in the reactor produces additional plutonium which, in turn, is used with more U-238 to fuel more LMFBRs. Because more than 99 percent of natural uranium is U-238, breeder technology would dramatically extend uranium reserves. The LMFBR uses liquid sodium as a coolant, generates higher temperatures than the light-water reactor, and operates at 40 percent efficiency.

The breeder program is the federal government's largest energy research and development project; it consumes more than 25 percent of ERDA's research and development budget. Estimates of the cost of the demonstration plant have escalated from the $400- to $500-million range in 1974 to $1.9 billion today. The costs of the overall program through 2020 were estimated in 1971 to be $3.9 billion; now, they are expected to be $12 billion.[4]

Although opposition to the LMFBR appears to be growing, the program commanded powerful support from the Ford administration, Congress, and industry. More than 350 utility companies have pledged $250 million to an LMFBR demonstration plant project being constructed at Oak Ridge, Tenn. This plant is scheduled to be in successful operation by 1983; commercial introduction of the LMFBR is expected by the late 1980s, with perhaps as many as 186 commercial breeder reactors operating by the year 2000.[5] Most observers feel, however, that this schedule will slip and could be delayed considerably if unexpected technical difficulties are encountered or federal research funds are cut.

As with the HTGR, arguments over the technology for the breeder reactor (as distinct from arguments over its safety, which are discussed in the following chapters) abound. The points can be summarized as follows:[6]

Arguments for industry and ERDA argue that development of breeder reactors is required to conserve uranium reserves and assure an adequate and economic energy supply in the future. In June 1971 President Nixon committed the United States to successful development of LMFBR technology by 1980 as "our best hope today for meeting the nation's growing demand for economical clean energy."[7] ERDA officials claim that if the breeding process is used, "uranium can provide us with fuel for centuries to come."[8] Others argue that because the breeder operates at a lower pressure, it presents less of a safety hazard in the event of an unforeseen release of radioactive steam than does the light-water reactor. Finally, some supporters of the breeder maintain that because it can use recycled plutonium as its fuel, it provides one answer to the problem of disposing of plutonium wastes.

Arguments against: Critics believe that much of the evidence presented in support of the LMFBR is flawed and that breeder technology involves hazards that far outweigh any potential economic gains. Many critics dispute the ERDA's contention that the rapid development of the LMFBR is essential to assure low-cost fuel for the nuclear power industry. They argue that the size of U.S. uranium reserves has never been determined accurately, and that they may be far greater than the ERDA estimates. Others, pointing to the spiraling projected costs of the LMFBR, assert that ERDA estimates of the economic benefits to be derived from the breeder program are either overstated or unsupportable. Critics also say that ERDA's cost-benefit analysis is based on questionable assumptions in such critical areas as nuclear power plant capital costs, fuel cycle costs, performance characteristics of the LMFBR, and electrical energy demands. In addition, critics say the high proportion of research and development funds devoted to the breeder is clear evidence that the government is giving less consideration to nonnuclear alternatives to the LMFBR. They feel this could lead to commercial use of the LMFBR despite potential advantages of other energy sources.

NOTES

1. Telephone conversation with General Atomic, December 1976.

2. For a description of light-water reactors, see Atomic Energy Commission, "The Nuclear Industry 1974," WASH 1174–74 (Washington, D.C., 1974), p. 23; Ralph Lapp, "Nuclear Salvation or Nuclear Folly?" New York *Times Magazine,* February 10, 1974, p. 12; Fred H. Schmidt and David Bodansky, *The Fight Over Nuclear Power* (San Francisco: Albion Publishing, 1976).

3. Energy Research and Development Administration, *A National Plan for Energy Research, Development and Demonstration: Creating Energy Choices for the Future,* 1976, Vol. 2, Program Implementation (Washington, D. C.: 1976), p. 281.

4. Ibid., Vol. 1, p. 37; also telephone conversation with ERDA, December 1976.

5. Ibid., Vol. 2, pp. 259 ff.; also, Joint Economic Committee, Hearings on the Fast Breeder Reactor Program, (Washington, D.C.: April 30 and May 8, 1975), p. 7.

6. For additional information on the LMFBR, see Brian G. Chow, *The Liquid Metal Fast Breeder Reactor* (Washington, D.C.: American Enterprise Institute for Public Policy Research, 1975); Thomas B. Cochrane, *The Liquid Metal Fast Breeder Reactor* (Baltimore: Johns Hopkins University Press for Resources for the Future, 1974); Joint Economic Committee, op. cit.

7. President Richard Nixon, Energy speech to Congress, June 1971.

8. Telephone conversation with ERDA Office of Public Information, December 1976.

CHAPTER

7

ACCIDENTS AT NUCLEAR PLANTS: POTENTIAL AND PROBABILITY

A major component of the debate over nuclear power has been the safety with which nuclear plants can be operated. Concern about nuclear safety has inspired public opposition to plant construction and prompted increased regulatory activity. It also has led to delays in construction and shut-downs in operation, and in doing so, it has imposed new costs on the industry and contributed to uncertainties affecting the pace of nuclear growth.

Few people disagree that the potential for danger exists at nuclear power plants, from accidents that would result in a release of radioactive materials. But there is strong disagreement on the likelihood that major accidents will occur at nuclear plants, the potential danger from such accidents, or the efficacy of measures that are being taken or might be taken to ensure that accidents do not occur.

THE POTENTIAL FOR DANGER AT NUCLEAR PLANTS

Danger of Radiation Exposure

A 1,000-megawatt light-water reactor—the average size now being built—contains a great deal of radioactive material when in operation, including such known carcinogens as strontium 90, cesium 137, and iodine 131. Varying levels of radioactivity also are present at iodine points in the nuclear fuel cycle—in uranium mining and milling, fuel fabrication, fuel reprocessing, and waste disposal. All authorities agree that accidental release of a large quantity of radioactivity, either from a plant or elsewhere in the fuel cycle, could have catastrophic effects on human beings. High levels of radiation can cause almost immediate death. Lower dosages can cause cancer of the thyroid, lung, or bone or lead to genetic damage. The severity of damage is dependent primarily on quantity and duration of exposure.

Disagreement exists, however, about the minimum amount of exposure likely to caùse damage to human beings. The American Nuclear Society (ANS) estimates that the average American is exposed to about 148 millirems of radiation each year. The exposure is from the following sources:[1]

Cosmic radiation at sea level	44 millirems
Rocks, soil, and building materials	55
Internal (inside body—what you eat, drink, or breathe)	25
Fallout (weapons tests)	4
Nuclear power	0.003
Medical diagnostic X rays	20
Total	148

The ANS estimates that humans are exposed to additional radiation during jet travel at a rate of about 2 millirems per 3,000-mile flight and from natural cosmic radiation, depending on altitude of about 1 millirem for every 100 feet above sea level. Thus, inhabitants of Denver may be exposed to 52 millirems more than residents of Los Angeles each year. ANS says that someone living at the boundary of a nuclear power plant would be exposed to 4.8 millirems a year if the plant were operating 24 hours a day, and anyone living more than five miles away from the plant would receive no increased exposure.[2]

The National Council on Radiation Protection and Measurements states that no person should be exposed to more than 500 millirems a year of man-made radiation (not including radiation in connection with medical examinations).[3] The American Nuclear Society estimates that exposure of 1 million people to a dose level of 1,000 millirems might produce 100 cases of leukemia.[4]

In recognition of the hazards posed by radioactivity, the Nuclear Regulatory Commission (NRC) has set standards and established procedures to ensure that emissions levels are "as low as reasonably achievable." What is reasonably achievable will be determined by a specific cost-benefit analysis to be performed by the company applying for a construction permit.[5] In numerical terms, the NRC operation guidelines recommend designs that would limit the radioactivity in effluents from nuclear power plants to levels that would keep the radiation exposures of persons living near the plants to less than 5 millirems a year. The NRC also restricts exposure of persons working with radioactive materials to 5,000 millirems a year.[6] Some critics argue these levels are too high, and some industry representatives assert the NRC standards are unnecessarily conservative.

A major cause of the disagreement about acceptable levels of exposure to radiation is the difficulty of determining whether any given case of illness is a result of radiation exposure. "There is no way of distinguishing between leukemia due to radiation and that arising from other sources," British radiologist Patricia Lindop writes.[7] Data have been extrapolated from experiments involving high levels of exposure to animals, but they are not sufficient to allow scientists to describe a definite relationship between level of radiation and incidence of cancer, and thus the questioning of standards has continued. A statement made in 1966 by the International Commission on Radiological Protection remains a valid summary of the problem. Although the limits adopted by the AEC provided reasonable latitude for expansion of atomic energy, the International Commission said, "it should be emphasized that the limit may not in fact represent a proper balance between possible harm and probable benefit, because of the uncertainty in assessing the risks and the benefits that would justify the exposure."[8]

Loss-of-Coolant Accident

The major fear in connection with reactor safety is not of low-level radiation, but of high levels of exposure as the result of a serious accident. Henry Kendall, professor of physics at MIT and a leading spokesman for the Union of Concerned Scientists—a group critical of nuclear power—argues that a major accident at a nuclear power plant involving failure of the plant's safety systems could result in release of nuclear radioactivity with widespread impact. "Consider, for example," Kendall says, "that 20 percent of a reactor's radioactive material is gaseous in normal circumstances, and, if released to the environment in one way or another, could be swept along by the winds for many tens of miles to expose people outside the reactor site boundaries to what could be lethal amounts of radioactivity. The lethal distance may approach 100 miles."[9]

The kind of accident most experts postulate could produce such a release would be a loss-of-coolant accident initiated by a rupture of a pipe in the primary cooling system. The water in the system is under high pressure, and a large pipe break would lead to rapid expulsion of the water out of the reactor core and into the containment vessel. In some cases this loss of coolant, or "blowdown," could take less than one minute.

The loss of water from the core would lead to an automatic shutdown, or "scram," of the reactor, but residual radioactive decay from the fuel rods would continue to generate heat. At shutdown, heat from residual decay is equal to 7 percent of the total heat generated by a plant in operation (3,330 megawatts of thermal energy in a 1,000-megawatt plant). It drops to 3 percent after 100 seconds and to less than 1 percent after two hours.[10] Unless additional cooling water is supplied to the core, the zirconium cladding around the fuel pellets is likely to deteriorate, and the fuel pellets will melt into a molten mass that might be hot enough to burn its way through the bottom of the reactor vessel, through the containment building and into the ground below. Where it would stop is not really known, a phenonemon that has been dubbed "the China syndrome."

In addition, it is possible that radioactive gases released from the fuel rods would escape from the pressure vessel along with the cooling water, and that the hot radioactive materials would cause a steam or chemical explosion that would burst the containment building, releasing the gases into the atmosphere.

To guard against the possibility of a loss-of-coolant accident, nuclear plants are constructed with what the industry describes as three levels of safety in a "defense-in-depth" or multiple-barrier system. The first level of safety results from conservative design and quality assurance for materials, design, and construction. The second level is based on redundant back-up safety systems, some of which are designed to shut down the reactor in case of emergency. As a third line of defense, the NRC requires that all light-water reactors be equipped with an emergency core-cooling system designed to flood the reactor core with water in the event of a failure in the primary system. The emergency system is supposed to begin functioning quickly enough after any loss of coolant to prevent the cladding on the fuel rods from rupturing as the heat builds up. It also is designed to cool the radioactive materials sufficiently to safeguard the containment building from destruction or rupture.

No major loss of coolant has ever occurred. In the few cases in which portions of the emergency system have been required, they have worked properly.

Industry and NRC officials express confidence that the emergency system will work. Their views are based on extensive research including computer modeling and semiscale testing used to predict the ability of the system to respond in case of a loss of coolant. There has been some debate, however, about the extent to which the tests that have been conducted are meaningful. In six tests on semiscale models carried out in 1970 by Idaho Nuclear Corporation under contract to the AEC, water in the emergency system did not reach the portion of the model designed to represent the reactor core. Critics have viewed the tests as a demonstration of the inadequacy of the emergency core-cooling system.[11] Industry and AEC officials have stated, however, that the true intent of the tests was not to simulate the working of the emergency system, but to test the effectiveness of the computer codes and that the design of the test models was not physically representative of conditions in full-sized reactors.[12]

As a result of the tests, the AEC conducted an extensive review and proposed new criteria for evaluating the performance of the emergency core-cooling system in nuclear plants. Critics, led by members of the Union of Concerned Scientists, maintained that there were flaws in the proposed criteria. In 1972 and 1973 the AEC held extensive rule-making hearings—involving 125 days of testimony and 55,000 pages of transcripts and documentation—on proper design criteria. Following the hearings, the AEC issued new regulations specifying that the performance of the emergency systems is to be evaluated according to new and more conservative criteria that take account of certain phenomena, such as fuel rod swelling and rupture and changes in the geometry of the reactor core or in the heat transfer coefficient—which critics had argued were likely to occur in the event of a loss-of-coolant accident and which had not been included in earlier models. NRC officials say the new criteria take into account the results of the earlier semiscale tests; industry officials argue that the new criteria are conservative.

The NRC has been conducting semiscale tests on a continuing basis and officials report that test results are "coming very close to predictions."[13] Tests involving an actual reactor core will begin in 1977. Indutry officials say that tests are being run to evaluate analytical methods used to predict a loss-of-coolant accident. NRC officials say that design differences on the experimental test models make it extremely difficult to consider the tests demonstrative of the actual behavior of large operating light-water reactors.

Operating History

The Nuclear Regulatory Commission requires nuclear plants and facilities "to report incidents or events that involve a variance from the regulations such as personnel overexposures, radioactive material releases above prescribed limits, and malfunctions of safety-related equipment."[14] The NRC investigates all reported events, and its findings are circulated "to the industry, the public, and other interested groups."[15] They are also available through the NRC's 122 public document rooms across the country. In addition, the NRC reports quarterly to Congress on any of these incidents that could be considered an "abnormal occurrence," defined as "an unscheduled incident or event which the commission determines is significant from the standpoint of public health and safety."[16]

In 1975 the NRC received reports on 1,650 events. Of these, the commission determined that 22 should be considered abnormal occurrences. None of the occurrences, the NRC reported, "affected the public health and safety; adequate levels of protection were available in each event, . . . there were no off-site exposures to radiation caused by these events, and there were no off-site releases of radioactive material that exceeded regulatory limits."[17]

Events may occur singly or at more than one plant—in which case they may be labeled a generic problem. Some of the areas in which plants have suffered recurring difficulties, although not usually of a serious enough nature to be ruled an abnormal occurrence under current NRC criteria, are as follows:

Valves: In 1973, valves were blamed for more abnormal occurrences than any other component; valve mechanism malfunctions recur in every nuclear plant, in every size, type, and brand of valve.[18] An AEC official stated in 1973 that "it appears highly possible that the incidence of valve malfunction actually is increasing." Valve failures in areas of the plants unrelated to the nuclear reactor have resulted in the deaths of two technicians; other valve failures have led to the exposing of a reactor core and have rendered high-pressure coolant systems inoperable.

Instrument switches: AEC officials told IRRC in 1974 that instrument switch set points are "constantly beyond required safety limits," and because of the frequency with which switches have been found to be out of calibration, the AEC suggested weekly inspections for some systems. In 1972, 70 percent of the operating nuclear plants had safety systems with defective switches on their motorized valve operators, rendering many safety relief valves inoperable. Misaligned, loose, and undersized switches have also been responsible for rendering high-pressure coolant systems inoperable.

Pipes: A number of reported events have had to do with pipes, pipe fittings, hangers, and supports. Several have affected the primary cooling system. Leaks detected in a pipe in the cooling systems of three boiling-water reactors led the NRC to ask 23 utilities to stop their reactor operations in order to inspect for cracks in their bypass lines, in case the cracks were generic to the pipe design.

Fuel: Problems have developed as a result of shrinkage of fuel pellets within the fuel rods. The shrinkage, or "densification," of fuel occurred in a number of pressurized-water reactor plants, leading to a collapsing of the fuel rods, an increase of fuel temperature, and the potential of leakage of fissionable materials from the fuel assemblies. The AEC temporarily reduced ("derated") the permissible level of power generation of a number of plants by 5 to 20 percent, until fuel rods could be replaced. Both the NRC and Westinghouse say that the difficulties of the densification have been eliminated through design changes.

Human error: A large number of the reported events have been attributed to human error. Specific incidents of human fallibility include insertion of tubes in the control rods upside down, incorrect installation of valves, improper attending of control rooms, and connection of waste disposal lines to a water fountain.

One of the most serious accidents ever reported at a commercial reactor occurred at TVA's Brown Ferry plant in March 1975. An engineering aide using a candle to check for air leaks set fire to polyurethane materials in a cable room. The $150-million fire did extensive damage to cables that were critical for control of the reactor and for operation of its emergency safety equipment. Some TVA engineers are reported to have said that a potentially catastrophic release of radiation was avoided "by sheer luck."[19] Industry representatives disagree that the Browns Ferry accident was a near catastrophe, and they argue that the fact that the accident was not more serious demonstrated the security of the safety back-up systems at nuclear plants.[20]

THE PROBABILITY OF ACCIDENTS AT NUCLEAR PLANTS

Differing interpretations of the significance of past problems with nuclear reactors and differing views about the reliability of emergency core-cooling systems have created sharp disputes over how likely it is that a loss-of-coolant accident will occur and what damages such an accident is likely to cause. Critics argue that the operating history is so replete with examples of errors and defects, and testing of the emergency core-cooling system has been so inadequate, that convincing evidence that nuclear power plants are safe should be developed before nuclear power is expanded further. Supporters of nuclear power respond by citing the study prepared under the direction of MIT professor Norman Rasmussen, and issued on behalf of the NRC in 1975, which concluded that a major accident—one involving the deaths of 3,300 or more people—is likely to occur only once in a billion reactor-years of operation, and that the damage from such an accident would be much less than had been predicted in studies sponsored earlier by the AEC.[21] Moreover, they defend the testing on the emergency core-cooling system and say that parts used in particular plants are being tested continuously, on a day-to-day basis.

The debate over accident probability and over the adequacy of safety research is important not only because of the damages a major accident would cause, but because most observers agree that a major accident at any nuclear power plant would probably evoke a sharp reaction that could include shutting down all operating plants. To appreciate the dispute, which now focuses on the Rasmussen report, it is useful also to review two earlier reports that evaluated the probable effects of a major accident and have shaped the character of the debate between critics and supporters of nuclear power.

The 1957 Brookhaven Report

In 1957 the Brookhaven National Laboratory completed a study, financed by the AEC, estimating the "theoretical possibilities and consequences of major accidents in large nuclear power plants."[22] The study concluded that the likelihood of an accident that would result in a major release of radioactive fission products outside the reactor containment ranged from 1 in 100,000 to 1 in 1 billion per reactor year. Assuming the worst possible combination of circumstances—including a climatological inversion, the breach of all containment barriers, absence of safety systems, and inability to evacuate the population—the study estimated that a major release could result in 3,400 fatalities, 43,000 injuries, $7 billion in property damage, and 150,000 square miles of contaminated land. In the introduction to the study, its authors counseled: "It should be emphasized that these numbers have no demonstrable basis in fact and have no validity of application beyond a reflection of the degree of their confidence in the low likelihood of occurrence of such reactor accident."[23]

Critics often have referred to the Brookhaven report in citing the dangers posed by nuclear power plants. They have argued that the plants analyzed in the report were about one-seventh the size of those operating today and that the study assumed that plants were located farther from population centers—30 miles—than has always been the case.

Supporters of nuclear power argue that the Brookhaven study is outdated. Then-AEC Chairman Dixy Lee Ray stated during 1974: "That 1957 report doesn't apply today. For one thing, reactors today feature a whole panoply of safeguards built into basic design so as to prevent accidents and to mitigate their consequences should one occur. Wash-740 [Brookhaven] assumed conditions that were so extreme that they would be virtually impossible to accomplish even if there was some unimaginable reason to try to do so."[24]

The 1965 Study

In 1964 and 1965 the AEC undertook to update the Brookhaven report. The commission prepared what it calls "an unfinished draft report" that concluded: "Assuming the same kind of hypothetical accidents as those in the 1957 study, the theoretically calculated damages would not be less, and under some circumstances would be substantially more, than the consequences reported in the earlier [Brookhaven] study."[25] In one extreme case, using what it called "grossly unrealistic assumptions" such as a major release of radioactivity and a simultaneous failure of all safety systems at a reactor located in the center of a city, the report estimated an accident could cause 45,000 fatalities, $17 billion in damages, and contamination of an area equal in size to the state of Pennsylvania.[26]

The 1965 study was not released until 1973, after portions of it had been obtained by nuclear power critics who threatened suit to force disclosure of the full document. The AEC maintained in 1973 that the 1965 study was never completed, that it was only released to "help the public to obtain a clearer understanding of the study undertaken some eight years ago of the probabilities and consequences of theoretical reactor accidents which have little resemblance to real situations." AEC officials argued that "meaningful information will be obtained in the [then in process] Rasmussen study."[27]

The Rasmussen Study

The Rasmussen study of reactor safety, officially titled "An Assessment of Accident Risks in U.S. Commercial Nuclear Power Plants," appeared in draft form in August 1974 and was issued in final form in October 1975. The objective of the study was to provide a quantitative assessment of the risks to public health and safety from potential accidents at nuclear plants. The study was directed by Norman Rasmussen, head of the department of nuclear engineering at MIT, and it was performed by a group of experts including employees of the federal nuclear and energy agencies and laboratories, independent consultants, and employees of private companies.

The report issued by the Rasmussen group adopted a system of event-tree/fault-tree analysis, developed originally by the Defense Department and the National Aeronautics and Space Administration, to estimate the possibility of reactor accidents. The analytic system used involved identifying all sequences of events that might occur if any given piece of equipment should fail, and continuing this analysis through to describe sequences in which a major accident would occur. Then, the probability of failure of any given piece of equipment involved in such a sequence was estimated, after which the probability of a chain of failures leading to a major accident could be computed.

The Rasmussen study concluded that the possible consequences of potential reactor accidents are predicted to be no larger and in many cases much smaller than people have been led to believe by previous studies which deliberately maximized estimates of these consequences. "The likelihood of reactor accidents is much smaller than that of many non-nuclear accidents having similar consequences. All non-nuclear accidents examined in this study, including fires, explosions, toxic chemical releases, dam failures, airplane crashes, earthquakes hurricanes, and tornadoes, are much more likely to occur and can have consequences comparable to, or larger than, those of nuclear accidents."[28]

Following the publication of the 1974 draft, the NRC requested comments from a number of organizations and individuals concerned about nuclear power. Ninety individuals and groups responded with some 1,800 pages of comments, and to a varying degree, the final report, issued in October 1975, reflects these comments.

The final version of the Rasmussen report draws several conclusions. It finds that the probability of a core melt—1 in 20,000—is greater than expected, but the likelihood of a fatality from a core melt—less than 1 in 20,000—is lower than had been anticipated. The study reports that "the consequences of reactor accidents are smaller than many people had believed." A serious accident, resulting in more than 10 deaths, is likely to occur once in 3 million reactor-years of operation. A most serious accident, one likely to occur once in a billion reactor-years, would kill 3,300 people, and cause latent cancers in 45,000. It would do an estimated $14 billion in property damage. The fatality risk to members of the public from 100 nuclear plants is lower than from hurricanes, tornadoes, or even falling meteors. It is, the study says, far less risk than from an automobile accident.[29]

THE CONTROVERSY OVER NUCLEAR PLANT SAFETY

No one disputes that nuclear power plants contain large quantities of dangerous radioactive materials that would cause significant losses of life and property if they were released into the external environment. Because of this danger, ensuring that nuclear power plants are safe has been a major concern of industry, government, and critics of nuclear power. Not even the most ardent supporters maintain that absolute safety is obtainable. The debate, therefore, centers on questions of what the risks are, how the risks can be measured, what is the acceptable balance between risk and safety, and how safe is safe enough. These are the major arguments raised on both sides of these questions:

Interpreting the Record

Opponents of nuclear power say the performance record of nuclear reactors now in operation is replete with design, fabrication, installation, operation, maintenance, and human errors affecting plant safety. Proliferation of nuclear plants, they say, will unduly increase the risk that similar errors will result in a serious accident. Proponents argue that the record of operating reactors provides evidence of the safety of nuclear power, not cause for concern. They say a combination of stringent regulations, safeguards, and contingency systems has enabled utilities to identify and deal safely with a number of minor problems; after almost 300 reactor-years of operation, there has not been a single nuclear-related accident affecting public health.

Critics

Critics claim that the number of events or incidents being reported demonstrates that the industry, despite its extensive efforts at quality assurance, is not able to plan for all contingencies and thus is not able to provide the systems and equipment required to guarantee public safety. In support of their claim, critics often cite AEC or NRC sources. One AEC draft report stated in 1973 that safety deficiencies "were besieging nuclear plants now operating or under construction,"[30] and a second commented that "the number of defects, equipment malfunctions, or failure events that have been encountered during construction, preoperational testing and routine nuclear power operation to date has been large, attesting to the fact that there is considerable room for improvement in practice, if not in philosophy."[31]

Critics give a number of reasons for these problems. They argue that the NRC is overly theoretical in its approach to safety or too vague in its guidelines. They say that the commission, often for commercial reasons, will allow old plants to operate without meeting new standards or will fail to halt operation of new plants that do not meet safety criteria. Or they say that utilities are simply not prepared to handle the complications of operating a sophisticated nuclear technology.

The resignation of three engineers from GE and two from the NRC gave new impetus to these complaints in 1976. Gregory Minor, formerly of GE, testified before the Joint Committee that "the industry, with the concurrence of the NRC, has overemphasized the theoretical approach to design verification with insufficient prototype, laboratory or field test verification." The result, he said, "is an inadequate and unsafe design."[32] And in later testimony, the GE engineers and Robert Pollard of the NRC focused their criticism on specific design defects. The engineers from GE criticized aspects of the reactor core, the control rods, the reactor vessel, the primary containment, and the supporting systems.[33] Robert Pollard, in a report to the NRC after his resignation, expressed concerns about the location of certain valves, problems with turbines, and the placement of electrical equipment.[34] A consultant to the NRC, Keith Miller, has recently written to the NRC faulting the design and testing of the emergency core-cooling system.[35] The fire at Browns Ferry provided critics with a clear focus for their concerns, and a number cite the incident as an example of the inadequacy of the NRC safety program. They quote an NRC report on the incident which states that "although some attention was paid to mitigating the consequences of fires, the NRC program in fire prevention and control was essentially zero." The NRC described its investigation as revealing "lapses in quality assurance in design, construction and operation."[36] Nevertheless, critics point out, the plant was licensed and allowed to operate.

In part, nuclear opponents say, the problem lies with utilities' inexperience. An AEC report commented in 1973 that "utility management, for the most part, has been slow to recognize the distinction between the organization and controls required to operate a nuclear power plant and the traditional controls employed in operating fossil fuel plants."[37] As mentioned earlier, opponents of nuclear power have criticized both utilities and reactor manufacturers for moving the industry along too quickly and for buying and building ever larger reactors before they have been sufficiently tested. But problems with nuclear power, critics add, may also result from an unwillingness on the part of utilities to take safety regulations as seriously as they should. The critics quote an NRC inspector as saying after the Browns Ferry fire that the "NRC, quite candidly, is trying to ram quality control down TVA's throat."[38] The commission, he said, was encountering some resistance.

At the base of this resistance may be a utility's financial concerns. The commercial prospects of nuclear power, critics argue, may conflict with an affective safety program. "The tremendous cost, schedule and political pressures experienced make unbiased decisions, with true evaluations of the consequences, impossible to achieve," the three GE engineers said in testimony to the Joint Committee after they had left the company. "The primary focus of the [assessment] program has been to 'prove' the plants are safe enough for continued operation—not to openly assess their true safety. . . . It is unfortunate that the commercial and technical proprietary pressures of the business world also work to the detriment of the maximum achievement of safety."[39]

Even if industry and the NRC were able to correct all design defects and resolve conflicts between safety concerns and commercial interests, however, many critics question whether it will be possible to achieve what they would consider to be an adequate level of safety. They say that nuclear plants are subject to Murphy's law—whatever can go wrong, will go wrong—and that many of the problems occurring in nuclear plants do not result from lack of experience with nuclear technology. To support this contention, they point to a comment in the AEC's magazine *Nuclear Safety* which said:

> During design and operation reviews of nuclear facilities, people frequently say, "any potential failure or accident you can think of can be avoided by proper design and operation." We usually sagely tell each other that it is the things we do not think of that will obviously someday suddenly confront us with our future problems. However, most of the occurrences reported are in direct contrast with this premise. The causes (of most of the reported failures) have been discussed at many reactors, and plans have been made at practically every installation to forestall their happening. Yet the failures still occurred.[40]

Some of the problems may simply prove to be without solution, critics say. Hannes Alfven, a 1970 Nobel laureate in physics, comments that one cannot claim that safety problems have been solved simply by pointing to all the efforts made to solve them.[41] Perhaps the most serious problems, and the least easy to solve, opponents say, are those that arise from human error. Amory Lovins, a physicist working with Friends of the Earth, suggests that ensuring nuclear power plant safety may involve not only engineering problems, but "rather a wholly new type of problem that can be solved only by infallible people."[42] Henry Kendall described nuclear power plants in 1974 as representative of "a system designed by geniuses being run by idiots."[43]

Proponents

Industry disputes critics' claims that a significant amount of concern with the operational safety of nuclear power plants comes from persons who have technical experience with their operation. Westinghouse says, "we believe that the actual number of critics is small. . . . In our view, most knowledgeable individuals favor nuclear power."[44] General Atomic agrees, and a spokesman wrote to IRRC that he thought "the critics represent small but vocal groups that now appear to be converging under the various Ralph Nader organizations."[45] Industry spokesmen cite testimony by Henry Kendall that the Union of Concerned Scientists has only about a dozen active members, and they argue that criticism of nuclear power is often based more on political bias than on technical knowledge. A number of industry officials described the campaign against nuclear power as one in opposition to the system. "Their objective is to force a change in the political system and our life style," ran a typical comment, "and they are using the debate over nuclear power as a means to this end."

Industry and NRC officials disagree with critics' interpretations of the operating safety record of nuclear plants and with their criticism of the NRC's regulatory role. Former NRC Chairman William Anders testified that the "competence, capability and dedication of NRC staff members and the organizational procedures under which they operate provide a safety review and basis for decisions that are second to none."[46] Both industry and NRC officials agree with Anders's comment that "the record speaks for itself: reactors designed and operating with redundant safety systems, highly conservative safety margins, and subject to regulatory review, have resulted in no reactor accidents which have impacted on public safety."[47] One utility characterizes industry's record as flawless, and Philip Bray, general manager of GE's boiling-water reactor department, states, "I know of no other industry in which safety has received more attention both in and out of government than the nuclear power industry. I know of no industry in which the safety programs have been more successful."[48]

Nuclear supporters do not say, however, that the industry is without its flaws. Anders, in commenting on some of the questions raised by the three engineers who left GE, testified in 1976 that he wanted to acknowledge that "some of the concerns, though not new, are not without merit."[49] But industry officials argue that they are working to correct any deficiencies that do exist and, at the same time, to provide safety back-up systems that prevent them from having any impact on a plant's safe operation.

Industry spokesmen credit what they describe as extensive back-up systems with the successful identification and control of potentially troublesome events. "General Electric has always followed the defense-in-depth and 'multiple barriers' concepts," as required by the federal government and practiced by the entire industry, Bray comments, "to assure the maintenance of high levels of safety. General Electric has, in designing plants, provided extensive multiple protection even against extremely unlikely events, . . . even in the event of equipment failure or procedural error."[50] GE and Commonwealth Edison, the utility most dependent on nuclear energy, cite the discoveries of cracks in pipes in boiling-water reactors as evidence that the quality assurance and control systems can identify problems before they become serious threats to public health and safety. And some industry officials point to the accident at Browns Ferry as a demonstration of how effectively the defense-in-depth system works. Aubrey Wagner, chairman of TVA's board of directors, testified before the Joint Committee that "the Browns Ferry incident has demonstrated the soundness of the underlying defense-in-depth design philosophy. . . . In effect, the Browns Ferry fire was a test—although a most unwelcome one—of the ability of a nuclear power plant to shut down safely under very difficult and extreme conditions."[51]

Industry and government officials also take issue with assertions that the design criteria are vague. "While it is true that many design criteria in our regulations are expressed in relatively general terms, regulatory guides give more expression to our staff's view of how particular design criteria may be met," NRC chairman Anders testified in 1976. "In an area of still-developing technology, it is often wise to leave room for improvements and alternative ways of doing things so long as safety is assured."[52]

Anders also disputes assertions that the NRC has failed to enforce standards once they are set. "If we believe a plant must be derated or shut down to assure public safety, then we will require it; if more time is required to make a sound decision, we will take it regardless of the cost to the utility and its customers. . . . The charge of economic bias is simplistic."[53] Anders explained that the NRC's basic mission to assure public safety had to be "mindful" that unnecessary delays should be avoided, but he rejected the idea that the recognition of the impact of certain regulatory actions on utility costs or power needs seriously hampers the NRC's ability to make unbiased decisions governing safety.

Utilities also reject the charge that they bypass safety because of commercial considerations. Con Ed told the Joint Committee that it spent "in excess of $3 million to build an actual similator of our plants, the Indian Point 2 and 3 plants at the station, so that we could be sure to train our personnel exactly on a duplicate of plants they will be operating."[54]

Finally, industry spokesmen say that they recognize that human errors represent a serious threat to plant reliability, but they are quite sure that systems can be constructed which accommodate to human weaknesses. Cleveland Electric Illuminating describes nuclear plants as having "fail-safe protective devices which can compensate for human error, monitoring and control systems and multiple containment barriers of steel-lined reinforced concrete one inside the other."[55] GE's Philip Bray testified to the Joint Committee: "We design so as to provide large safety margins to more than accommodate any uncertainty regarding specific phenomena."[56]

The industry's commitment to safety, according to its spokesmen, is demonstrated by quality design and construction, extensive back-up systems and redundant controls, and quality assurance programs in production and operations. Plants are built with multiple barriers to contain radioactive materials—the zirconium cladding, the reactor pressure vessel, a reinforced concrete containment building—and with emergency systems in case of failure of primary systems. "We are not perfect, but we incorporate backups and conservatism all along the way so that, when failures do occur, they can be handled safely,"[57] says W. H. Arnold of Westinghouse. "Safety starts in the design." Design integrity, industry representatives say, is maintained through the most extensive quality assurance programs ever used by a commercial industry. The programs involve widespread reporting and documentation and are often administered by a separate staff. General Electric reports that it has one employee working on quality assurance for every four on production at its fuel fabrication plant.[58] Westinghouse representatives describe their commitment to safety as personal as well as institutional: "Our own personal and professional futures are involved."[59]

Industry representatives also reject contentions that growth has been too rapid. Bray told IRRC that GE was limited by its own internal committees to constructing or using parts that represented no more than a 10 percent variation from those that were already in use. Moreover, GE's hazards committee often set standards more rigorous than those set by the AEC, Bray said, maintaining that its safety and production standards were more conservative than necessary.[60]

Safety Research and the Emergency Core-Cooling System

Much of the contention between nuclear proponents and their critics has concerned the nature of the safety research conducted by industry, the AEC, and now the NRC. Critics fault safety research to date as incomplete and inadequate, and they argue that both the government and industry have misrepresented the results of safety research to the public. A specific focus of the critics' concern has been the emergency core-cooling system, the principal backup safety system that must operate in cases of a serious accident. Nuclear opponents say that it has never been properly tested. Proponents argue that although no full-scale test of the emergency system has been carried out, hundreds of tests conducted on the system's component parts have provided adequate evidence of its ability to operate in the unlikely event of a major accident.

Critics

The Union of Concerned Scientists argues that the AEC and NRC testing programs have been narrow in scope and that experimental results for a loss-of-coolant accident and response by the emergency core-cooling system (ECCS) cannot be predicted accurately. The computer programs designed to simulate a "loss" and the functioning of the ECCS are described as "inaccurate and unverified" by Carl Hocevar, a safety research engineer who did contract work for the AEC but resigned in 1974 to join the Union of Concerned Scientists.[61] Because of the faultiness of the codes Hocevar has said, "the safety of nuclear power plants has not been demonstrated."[62]

Critics also fault the testing program as too little, too late. They say that the NRC's loss-of-fluid tests are something of an afterthought, because more than 60 plants are already in operation. Moreover, they say that the results of the test—to be made on a 15- to 20-megawatt reactor—may not be applicable to modern reactors that are nearly 60 times larger.[63]

Critics' concerns about the ECCS are discussed in a recent independent study on reactor safety by the American Physical Society.[64] In a summary of the report, the authors write that while they have no reason to doubt that the system will work as planned, they also have "no comprehensive, thoroughly quantitative basis for evaluating ECCS performance, because of the inadequacies in the present data base and calculational codes."[65] Results from loss-of-fluid tests—seen as critical to an accurate assessment of the safety system—are five to eight years away, according to the study.Current computer programs may well not prove applicable to the larger and more sophisticated reactors, the study says, and its authors conclude that they doubt that "a complete quantitative evaluation of ECCS effectiveness can be achieved through the present program."[66]

On a related issue, David Comey of Business and Professional People in the Public Interest wrote to the AEC on August 22, 1974, challenging the operating efficiency of the diesel engines which are to power the emergency core-cooling system in case of a loss of off-site power coincident with a loss-of-coolant accident. Comey pointed out that a number of the diesels had a reliability factor of less than 92 percent—the AEC's reliability standard is 99 percent—and cited one with a reliability factor of 51 percent. He concluded: "Fortunately, although there have been well over a hundred diesel generator failures at operating nuclear plants in the last few years, none of them has occurred coincident with a major accident. . . . If, however, the diesel generators were to fail to start in the event of a major loss-of-coolant accident with a loss of off-site power, then the plant emergency core-cooling system would be without power, and a delay of even 120 seconds in getting the diesels started by manual means would probably lead to onset of a fuel core meltdown."[67] The AEC said it shared Comey's concern and was taking steps to deal with the problem.

Proponents

Supporters of nuclear power disagree with the critics' analysis, much of which they consider to be based on inadequate or outdated information. Hans Bethe, Nobel physicist, argues that Carl Hocevar's criticism of computer codes that are used to test the ECCS refers to codes that are no longer used. He agrees that the old codes were not conservative, but he says that stricter design criteria have been adopted and that "as a consequence, most people who had previously doubted the effectiveness of the ECCS changed their minds."[68]

Industry representatives reject the critics' contention that full-scale tests are required to assure that nuclear plants are sufficiently safe. "A full-scale test is much less demanding on the system being tested," W. H. Arnold of Westinghouse has stated, "than a well-devised series of component tests coupled with a good program of analytical predictions of the effect of system failures on components. In a full-scale test, one takes what one gets. One has little or no control of influences a particular component will experience. In individual tests, one can subject individual components to conditions two, three, or even ten times worse than they might receive in the full-scale test."[69]

Not only is such a test unnecessary, nuclear proponents say, but to ask that one be done is unreasonable. "Even if such a test (pushing a plant to a true-core melt and destruction) were done, the test would be met with the claim we tested it the wrong way," GE's Philip Bray comments. "We could say that more plants should be tested to destruction, thus leading to the wasteful expenditure of billions of dollars and, more importantly I think, the wasteful diversion of our technical talent towards areas of reactor design that don't add to over-all safety."[70]

Most industry and government officials think that the debate over the ECCS was ended by the AEC's rule-making hearings and the adoption of more conservative criteria. The ECCS, one Atomic Industrial Forum representative told IRRC, is "a non-issue."

Likelihood and Consequences of a Nuclear Plant Accident

A critical aspect of the debate over nuclear safety is the controversy over the likelihood and potential consequences of an accident at a nuclear plant. Critics say that neither the government nor industry have accurately assessed the probability that a serious accident could occur or the damages that such an accident could cause. Proponents counter that the Rasmussen report, completed in 1975, uses the most modern and sophisticated techniques of risk estimation available. Moreover, they say, its findings demonstrate the extremely low level of relative risk from operation of nuclear plants. Much of the debate over the risks and consequences of nuclear power development focuses on the adequacy of the Rasmussen report and the adequacy of its findings.

Critics

Critics do not accept the Rassmussen report as a conclusive assessment of nuclear safety. Milton Kamins, a Rand Corporation engineer, in an update of a critique of the Rasmussen report by the Sierra Club and the Union of Concerned Scientists, criticizes the report because: (1) the researchers "used an incorrect and thoroughly discredited methodology for predicting future reliability . . .; (2) the authors of the report have applied the chosen methodology incorrectly in a number of respects; (3) [the study] is vague and inadequate concerning the likelihood of human error . . . and is virtually silent on plausible external threats, in particular sabotage."[71] A number of critics, including the Union of Concerned Scientists and the Sierra Club, also contended that government and industry misrepresented the results of the study as evidence of nuclear power's safety.

Some critics were particularly harsh on the methods—event-tree/fault-tree analysis—used by Rasmussen in his study. Robert Augustine of the National Intervenors described the report as a product of "guesses multiplied by guesses—sheer speculation." The processes involved in a meltdown and release of fission products during a reactor accident are largely unstudied and simply unknown, he argued.[72] Joel Primack, professor of physics at the University of California at Santa Cruz, describes the problem of determining probabilities of accidents at a nuclear plant as "gamblers' guesses." It is the problem, he said, of trying to determine the relative probability of unlikely events which have never occurred and about which no data exist.[73]

A joint study by experts for the Union of Concerned Scientists and the Sierra Club stated that the techniques used by Rasmussen "have consistently failed to make reliable estimates of accident probability" when used in the aerospace program, where they were developed. It also challenged the report's assumptions that all important design errors and accident sequences have been identified, arguing: "The experience in the aerospace program is instructive. Approximately 20 percent of the Apollo ground test failures and over 35 percent of the in-flight malfunctions and failures were system malfunctions that were not identified as 'credible' prior to their occurrence, despite a very thorough failure mode analysis made prior to the tests of in-flight trials. AEC records similarly indicate that completely unsuspected, and by no means trivial, safety problems are arising with frequency at the country's nuclear plants."[74]

The difficulty in predicting, critics say, is illustrated by the Browns Ferry fire. Rasmussen has testified that the likelihood of a fire of the sort that occurred at Browns Ferry is about 1 in 50, 000.[75] And yet, critics say, it happened. Henry Kendall, who presented the critics' report at a 1974 press conference, acknowledged that they could not suggest more effective ways of evaluating the probability of an accident. But, Kendall argued, the AEC and industry bear the burden of developing adequate analytic techniques. In light of the badly flawed performance record of existing plants, he contended, until better techniques are developed construction of additional nuclear power plants should be deferred.[76]

The Union of Concerned Scientists/Sierra Club study team also argued in a summary of its report that the Rasmussen study is flawed by "a number of weaknesses" in its assumptions. The team said Rasmussen underestimates by a factor of two the amount of radioactivity released in a core melting accident, and by a factor of one to six the potential damage from radioactive exposure. At the same time, the team stated, the Rasmussen study overestimates the possibility for prompt and effective evacuation in case of an accident and fails to take into account potential population growth around nuclear plants. Over all, it concluded, Rasmussen may have underestimated the consequences to human health of a major accident by a factor of 16.[77]

Some of these criticisms of the Rasmussen study have received support from the U.S. Environmental Protection Agency (EPA) and from the American Physical Society (APS). EPA officials have described the study as providing "the first credible assessment of the likelihood of major nuclear reactor accidents and their risks," but they say also that the study fails to evaluate adequately the consequences of a possible accident.[78] Deaths and injuries from a plant accident could be greater by a factor of two to ten than the Rasmussen study predicts, according to the EPA.[79]

The APS study concluded with an estimate of "substantially larger long-term consequences particularly concerning land damage/denial and possible latent cancers from exposures to individuals who live in areas which are contaminated." The authors of the study also stated that although they recognized the merits of event-tree and fault-tree analysis, in highlighting relative strengths and weaknesses of reactor systems, "based on our experience with problems of this nature involving very low probabilities, we do not now have confidence in the presently calculated absolute values of the probabilities of the various branches."[80]

Many observers have made clear that they consider that although the Rasmussen study is useful, it may also be misused. The Union of Concerned Scientists/Sierra Club study accepts that event-tree/fault-tree analysis "can be very helpful in making comparisons between diverse system designs," but it concludes that the methodology cannot "be employed as [Rasmussen] has done to determine absolute probability values for accident probabilities and to use these predictions as proof of the safety of nuclear plants."[81] And William Rowe of EPA says that while the study "provides a considerable advance in our understanding of the probability or likelihood of future accidents . . . it can make no prediction as to if and when an accident will occur."[82]

On the subject of the likelihood of an accident at the yet-to-operate breeder reactor plants, critics express concern that the breeder reactor involves a now-unproven technology which has safety problems that are at least an order of magnitude more severe than those of the light-water reactor. They say that development of the LMFBR should not go forward until all these problems are solved. Some critics fear that safety standards may be sacrificied to make the LMFBR economically attractive. Their major areas of concern about the operational safety of the LMFBR are: (1) the use of liquid sodium as a coolant introduces serious hazards, because liquid sodium is violently reactive with air or water, becomes intensely radioactive as it circulates around the reactor core, and is susceptible to bubbles or "voids" that might lead to a chain reaction in the fuel assembly and possible explosion; and (2) the presence of "fast" neutrons in the fuel core, which is operating at temperatures approaching the melting point of the encasing metals, greatly complicates the maintenance of the integrity of the fuel core. This integrity is critically important to avoid a core meltdown and possible release of radioactivity to the atmosphere.

Proponents

Government and industry representatives view the Rasmussen report as highly favorable to continued development of nuclear power. They say it demonstrates that the risk from nuclear power is small, and they defend the methods used as the most accurate and sophisticated available. Most importantly, they say, the study shows that nuclear power is far less dangerous than many day-to-day events, and that risks from nuclear plants are smaller than from alternative forms of energy.

Supporters of the study accept that criticism of the first draft was often relevant and useful, but they say that the final draft has made the corrections that were required. But on balance, the study authors say, the final version makes few major changes. "The probabilities predicted for potential accidents did not change significantly," Saul Levine, staff director of the study, told the House Interior and Insular Affairs Committee in June 1976. While the predicted consequences increased generally over the first report, "most of the values are within the predicted confidence bounds of ⅓ and 3 predicted in the draft report," with the exception of a calculation relating to cesium which altered estimates of latent cancer fatalities by a factor often. And the "conclusions of the draft report," Levine said, "that the risks involved in potential nuclear power plant accidents are very small compared to nonnuclear risks, did not change in the final report."[83]

Nuclear proponents dispute critics' objections to the methodology employed by the study's researchers. Levine argues that "there has been a broad and increasing acceptance by the informed scientific community of the study's methodology."[84] An Atomic Industrial Forum representative said that although the study uses techniques of analysis developed to assess risks of failure in the aerospace program, the techniques have been expanded and refined. Robert Szalay, AIF's manager of licensing and safety projects, told IRRC that the report estimates a range of probabilities for each possible event and sequence of events. "It is possible to have a pretty high confidence within each range," he said. "While no one can say that everything has been taken into consideration, the detail of examination of each component and the number of operating and design personnel interviewed by Rasmussen and asked 'What if?' provides assurance that the study has enveloped the events that are significant contributors." By making conservative assumptions, Szalay says, the study absorbs areas of uncertainty.[85]

A major contribution of the study, nuclear proponents say, is that it quantifies risks from nuclear and nonnuclear events and allows comparison of the risks from nuclear plants and those from alternative sources. They say, it demonstrates that the risk from nuclear plants is lower. Taking a pessimistic hypothesis, Richard Wilson, a physics professor at Harvard, states, "We can deduce that at the most there will be one cancer death in 10 years for every 1,000-megawatt nuclear power station. If we carry out a similar calculation for a coal-fired power station, we find that sulfate and particulates give an increase in deaths from bronchial ailments of from 50 to 100 per power station depending on location." Wilson goes on in his testimony to weigh the likelihood and consequences of a liquefied natural gas accident or a collapse of a hydroelectric dam against a nuclear accident and finds that the probabilities and the potential costs of either are larger than those estimated by the Rasmussen study for a nuclear plant.[86] "Compared with other accident risks that our society accepts, the risk from nuclear reactors is very small," Hans Bethe has stated.[87]

The study's authors dispute the contentions of critics that the study does not give proper consideration to sabotage, "common mode" failures, or the possibility of human error. "The important question is not whether all contributions have been included, but whether all significant contributions to risk have been covered," Levine has testified. The analysis indicated that "sabotage could not cause accidents larger than the largest predicted by the study," for example, "and so the authors did not try to predict the likelihood of sabotage."[88]

Both industry and government officials have quoted the Rasmussen report as indicating that nuclear plants are safe. "The report reinforces the commission's belief," Anders has said, "that a nuclear power plant designed, constructed and operated in accordance with NRC's comprehensive regulatory requirements provides adequate protection to public health and safety and the environment."[89]

In responding to charges that the liquid-metal fast-breeder reactor—not covered by the Rasmussen study—will pose greater risks, some officials argue to the contrary. Bethe maintains that not only will the breeder have all the safety features of the light-water reactor, but it will also have other aspects that increase its safety. He cites the low temperature of sodium relative to its boiling point and its high capacity to conduct heat—both features that improve response in case of an accident.[90]

The Price-Anderson Act

The Price-Anderson Act has become a focal point, at the legislative level, for debate on nuclear safety. Critics argue that convincing evidence of industry's uncertainty about reactor safety can be drawn from manufacturers' and utilities' insistence that their liability for an accident be limited by law and that federally subsidized insurance be provided to nuclear power plants, through extension of the Price-Anderson Act. And they have introduced bills and initiatives at the state level to end Price-Anderson supports. If nuclear power is as safe as the proponents say, critics argue, the risks should be small enough to be borne by the industry. By removing the liability, they contend, Price-Anderson may mitigate against the industry's taking sufficient responsibility on being adequately vigilant. Limiting liability removes a deterrent that is important to protecting the public against a negligent industry.

This point of view was represented by members of Congress who supported an amendment to end limits on liability. "This liability limitation might have been justified when this was an infant industry," Representative Jonathan Bingham (D-N.Y.) stated, "but there are big boys in the industry now and they should be able to stand on their own feet." Teno Roncalio (D-Wyo.) commented that "the nuclear industry is much too content with feeding at the government breast and simply does not want to assume its rightful responsibility as a mature, powerful and safe sector of our free market economy."[91]

Proponents

Proponents argue that the tremendous capital investment required for nuclear power, the newness of the technology, and size of the potential—if improbable—risk involved necessitates a unique kind of government participation in ensuring against an accident. They also say that the revised Price-Anderson Act calls for decreasing participation of government, with private industry bearing an increasing share of the responsibility.

Industry officials contend that, to the contrary, Price-Anderson provides a degree of protection that would not be available otherwise. Nuclear insurance representative Joe Marrone explains that without Price-Anderson, recovery of damages could be slow and difficult in the case of a nuclear accident. Proving causal connection between an accident and an injury would be difficult because, for example, leukemia caused by radiation is no different from leukemia caused by other sources, and latent damage may not appear for years. Price-Anderson provides for immediate compensation of claims, which are then credited against final settlements. It also sets aside funds for latent damages.[92]

Industry officials admit it is unlikely that private insurance companies would be willing to provide coverage equal to that available under Price-Anderson. However, a study prepared for the Atomic Industrial Forum by Columbia University maintains that this unwillingness does not necessarily reflect safety considerations: "While an activity might be fully insurable on the basis of safety considerations, it may be uninsurable to the full extent of the largest potential loss because of the size limitations of the insurance industry itself."[93]

The Columbia study argues that industry's desire to avoid full liability is fully "consistent with a belief that there is an exceedingly remote possibility that an accident will occur, but that liability for the full consequences of that accident would be disastrous."[94] Vendors of individual component parts, utilities, or even insurance companies could not absorb the consequences of such an accident, and Price-Anderson provides a method of sharing responsibility, the study says. "The position of industry in seeking to avoid the potential of full liability should be viewed as a reasonable business judgment."[95]

Industry officials support this position. Westinghouse states:

> It must be kept in mind that the traditional approach to public health in a new industry is the early establishment of a balance between actual experience and social acceptability. The insurance industry normally utilizes a process of actuarial determination of statistical performance to determine its economic approach to insurance coverage. In any field where the actuarial data do not exist, the normal functioning of private insurance companies becomes financially hazardous. Under such circumstances, when the furthering of the national interest would be inhibited by the absence of insurance coverage, the government has traditionally stepped in to provide such

> coverage. Such programs that have been implemented in the past include crop insurance, bank deposit insurance, savings and loan account insurance, maritime vessel insurance, mortgage insurance, maritime cargo wartime insurance, veterans' life insurance, unemployment insurance, and so forth.[96]

Industry representatives also discount critics' belief that the shared public-private liability under Price-Anderson will allow industry to be complacent. Any serious nuclear accident would involve financial losses far beyond liability coverage. Utilities are not insured against the costs of replacement electricity they must purchase if an accident shuts down one of their base-load plants. Such costs could run $100,000 to $200,000 a day. Moreover, the Columbia report asserts that an accident of catastrophic proportions would affect the entire industry, including utilities, vendors, architect-engineers, and suppliers, and bring its long-term viability into question. "In short," the report argues, "the impact of a catastrophic accident on the industry would be a disaster for the industry as well as the public. The argument that the limitation on the potential third-party liability of the industry discourages safety is thus not convincing."[97]

Moreover, industry spokesmen say, the industry has already made a substantial commitment to insurance for nuclear plants and will be increasing that commitment in the future. Officials of the Atomic Industrial Forum state that "private insurers provide for each nuclear power plant $140 million of property damage and $100 million of liability insurance, and they point out that claims for liability damage have been minimal, and that about two-thirds of liability damage premiums have been returned to the utilities.[98] General Electric says that private insurers have just agreed to increase their proportion of total coverage under the Price-Anderson Act. "It is particularly significant," GE stated, "[that the] increases come at a time when the conventional insurance industry is experiencing a period of poor underwriting results. The increase at this time expresses a rather unparalleled confidence in nuclear power by a major segment of the American financial community."[99]

NOTES

1. American Nuclear Society, "Nuclear Power and the Environment, Questions and Answers" (Hinsdale, Ill., 1976), p. 26.
2. Ibid.
3. National Council on Radiation Protection and Measurements, Basic Radiation Criteria, NCRP Report No. 39, January 1971.
4. American Nuclear Society, op. cit., p. 34.
5. Nuclear Regulatory Commission, "Licensing of Production and Utilization Facilities," 10 CFR 50, App. 1, Nuclear Regulation Reports, January 2, 1976, p. 8761.

6. Ibid., pp. 8760 ff.

7. Patricia J. Lindop, and J. Rotblat, "Radiation Pollution of the Environment," *Science and Public Affairs* (Bulletin of the Atomic Scientists), September 1971, p. 18.

8. International Commission on Radiological Protection, *Publication No. 8, 9,* (New York: Pergamon Press, 1966).

9. Quoted in Ralph Lapp, "Nuclear Salvation or Nuclear Folly?," New York *Times Magazine,* February 10, 1974, p. 64.

10. Westinghouse, letter to IRRC, March 18, 1976, p. 14.

11. Robert Gillette, "Nuclear Reactor Safety: At the AEC the Way of the Dissenter is Hard," *Science,* May 5, 1972.

12. Romano Salvatori, "Nuclear Power Plant Safety," speech before the Nuclear Power Pollution Subcommittee of the Committee on Marine Affairs of the Michigan House of Representatives, October 24, 1974, p. 14.

13. Interview with NRC officials, November 1976.

14. Nuclear Regulatory Commission, *Report to Congress on Abnormal Occurrences,* July-September 1975, p. 1.

15. Ibid.

16. Ibid., p. iii.

17. Ibid., October-December 1975, p. 1.

18. Atomic Energy Commission, "Summary of Abnormal Occurrences Reported to the AEC During 1973," May 1974, 002-05-001, p. 2.

19. Union of Concerned Scientists, "Browns Ferry: The Regulatory Failure" (Cambridge, Mass., June 10, 1976), p. 8.

20. See Joint Committee on Atomic Energy, "Browns Ferry Nuclear Plant Fire," September 16, 1975.

21. Nuclear Regulatory Commission, *Reactor Safety Study,* Executive Summary, Wash 1400, October 1975, p. 9.

22. Atomic Energy Commission, *Theoretical Possibilities and Consequences of Major Accidents in Large Nuclear Power Plants,* Wash 740 (the Brookhaven Report), March 1957.

23. Ibid., p. 6.

24. Dixy Lee Ray, Statement to the Press, press release, AEC, Washington D.C., November 16, 1974.

25. David Burnham, "AEC Files Show Effort to Conceal Safety Perils," New York *Times,* November 10, 1974, p. 1.

26. Ibid.

27. Atomic Energy Commission, press release, June 25, 1973.

28. Nuclear Regulatory Commission, *Reactor Safety Study, An Assessment of Accident Risks in U.S. Commercial Nuclear Power Plants* (Washington, D.C.: October 1975).

29. Ibid.

30. Union of Concerned Scientists, op. cit., p. 25.

31. Robert Gillette, "Nuclear Safety: AEC Report Makes the Best of It," *Science,* January 26, 1975.

32. Gregory Minor, Hearings before the Joint Committee on Atomic Energy, February 18, 1976, p. 9.

33. Ibid., pp. 494 ff.

34. Ibid., p. 594.

35. Keith Miller, "Memorandum to the Commissioners, USNRC," University of California at Berkeley, May 6, 1976.

36. Union of Concerned Scientists, op. cit.; also, David Burnham, "Inquiry on Fire at Biggest Nuclear Plant Finds Prevention Program Was 'Essentially Zero,' " New York *Times,* February 29, 1976, p. 27.

37. Atomic Energy Commission, *The Safety of Nuclear Power Reactors and Related Facilities,* Wash 1250 (Washington, D.C.: 1973).

38. Union of Concerned Scientists, op. cit., p. 25.

39. Dale Bridenbaugh, Richard Hubbard, Gregory Minor, Hearings before the Joint Committee on Atomic Energy, op. cit., pp. 539, 40, 41.

40. Atomic Energy Commission, *Nuclear Safety* 15 (1974).

41. Hannes Alfven, *Bulletin of the Atomic Scientists,* May 1972, quoted in "The Nuclear Power Issue: An Overview," Union of Concerned Scientists (Cambridge, Mass., n.d.).

42. Amory Lovens, "The Case for Long-Term Planning," *Bulletin of the Atomic Scientists,* June 1974, p. 44.

43. Henry Kendall, press conference to review the Rasmussen Report (Washington, D.C., November 24, 1974).

44. Letter to IRRC, December 17, 1974.

45. Letter to IRRC, January 1975.

46. William Anders, Hearings before the Joint Committee on Atomic Energy, February 23, 1976, p. 269.

47. Ibid., p. 270.

48. Philip Bray, Hearings before the Joint Committee on Atomic Energy, 1974, reprint from GE, p. 4.

49. Anders, op. cit., p. 267.

50. Bray, op. cit.

51. Aubrey Wagner, Hearings before the Joint Committee on Atomic Energy, September 16, 1976, p. 131.

52. Anders, op. cit., p. 267.

53. Ibid., p. 268.

54. John Conway, Hearings before the Joint Committee, February 23, 1976, p. 137.

55. Cleveland Illuminating Company, advertising materials sent to IRRC, 1974.

56. Bray, op. cit.

57. Interview with IRRC, 1974.

58. Ibid.

59. Ibid.

60. Ibid.

61. Carl Hocevar, "Nuclear Reactor Licensing—A Critique of the Computer Safety Prediction Methods," Union of Concerned Scientists, August 14, 1975, p. 4.

62. Carl Hocevar, *Hearings on Proposition 15,* California State Assembly, Vol. III, 1975, p. 57.

63. Joel Primack, *Hearings on Proposition 15,* California State Assembly, Vol. V, October 28, 1975, p. 21.

64. American Physical Society, *Report to the American Physical Society by the Study Group on Light-Water Reactor Safety, April 28, 1975,* National Science Foundation and the AEC.

65. Ibid., p. I–5.

66. Ibid, p. I–6.

67. David Comey, "Lack of Reliability of Emergency Diesel Generators at Operating Nuclear Plants," letter to L. Manning Muntzing, September 10, 1974, available from Business and Professional People in the Public Interest, Chicago.

68. Hans Bethe, *Hearings on Proposition 15,* California State Assembly, Vol. IV, October 22, 1975, p. 15.

69. W. H. Arnold, in ibid., p. 61.

70. Bray, op. cit., p. 98.

71. Milton Kamins, "A Reliability Review of the Reactor Safety Study," Rand Paper Series, April 1975, p. 2.

72. Robert Augustine, "AEC Finds Safety in Numbers," *Environmental Action,* October 12, 1974, p. 3.

73. Primack, op. cit., pp. 48, 67.

74. Union of Concerned Scientists/Sierra Club, "Preliminary Review of the AEC Reactor Safety Study" (Cambridge, Mass., November 1974), p. 19.

75. Norman Rasmussen, Hearings before the House Committee on Interior and Insular Affairs, February 26, 1976, p. 107.

76. Kendall, op. cit.

77. Union of Concerned Scientists/Sierra Club, op. cit.

78. William D. Rowe, testimony before the House Committee on Interior and Insular Affairs, June 11, 1976, p. 5.

79. Ibid., p. 6.

80. American Physical Society, op. cit.

81. Union of Concerned Scientists/Sierra Club, Press release, op. cit., Washington, D.C., November 12, 1974.

82. Rowe, op. cit., p. 2.

83. Saul Levine, testimony before the House Committee on Interior and Insular Affairs, June 11, 1976.

84. Ibid.

85. Interview with IRRC, 1974.

86. Richard Wilson, *Hearings on Proposition 15,* California State Assembly, Vol. V, October 23, 1975, pp. 3 ff.

87. Hans Bethe, "The Necessity of Fission Power," *Scientific American,* January 1976.

88. Levine, op. cit., pp. 7 ff.

89. Ibid., p. 22.

90. Bethe, "The Necessity of Fission Power," op. cit., p. 6.

91. Jonathan Bingham, and Teno Roncalio, reported by Elder Witt, "Nuclear Insurance Program," December 13, 1975, p. 2694.

92. Telephone conversation with IRRC, 1974.

93. Laurie Rochett, "Issues of Financial Protection in Nuclear Activities," Legislative Drafting Research Fund, Columbia University, December 21, 1973, p. 3–14.

94. Ibid., p. 4–3.

95. Ibid.

96. Salvatori, op. cit.

97. Rochett, op. cit., p. 4–2.

98. Westinghouse, comments to IRRC, December 17, 1974.

99. GE, letter to IRRC, January 10, 1975.

CHAPTER

8

SAFEGUARDING RADIOACTIVE MATERIALS

Many observers believe that safeguarding the products and by-products of nuclear waste—some of which are radioactive and some of which can be used to make nuclear weapons—pose problems far more serious than those raised by nuclear plant safety. They question whether safeguards can be devised which can prevent nations or terrorist groups from misusing the radioactive materials produced by a nuclear plant, particularly plutonium, to manufacture bombs or threats of bombs, and they also are concerned whether it is possible to develop safe and secure methods for permanent storage of highly dangerous radioactive wastes, some of which need to be isolated from the environment for hundreds of thousands of years.

THE PLUTONIUM PROBLEM

In the view of some, the possibility that radioactive materials during the nuclear fuel cycle will be used for military or terrorist purposes represents the most significant threat to public safety posed by development of nuclear power. Their overall concern is whether adequate safeguards exist, or could be developed, to prevent potential misuse of such materials. The greatest dangers, they believe, arise in connection with plutonium.

What Plutonium Is

Plutonium has been described by its discoverer, former AEC Chairman Glenn Seaborg, as one of the most toxic elements known to man. It is extremely carcinogenic; one millionth of a gram has been shown to cause cancer in animals.[1] More important, perhaps, is that physicists estimate that a crude "atomic bomb" could be constructed with about four kilograms of plutonium —about one-fiftieth of the amount of plutonium produced each year by a 1,000-megawatt nuclear power plant.[2]

Each 1,000-megawatt nuclear power plant produces, together with other materials in its spent fuel, about 200 kilograms of plutonium a year. By 1980, observers estimate, annual plutonium production from light-water reactors will be 27,000 kilograms; by the year 2000 annual production may grow to 385,000 kilograms.[3] Furthermore, if breeder reactors come into common usage, the production of plutonium will increase immensely. A breeder reactor will contain a ton of plutonium and will produce seven times as much plutonium as a light-water reactor of equal generating capacity. An energy economy in which breeder reactors play a major role in power generation and fuel creation would have nearly three times as much plutonium in circulation —more than 1 million kilograms of plutonium a year—as one involving only light-water reactors using recycled plutonium.[4]

The Plutonium Economy

The threat of plutonium misuse is small now because very little plutonium is available in a form that would make it susceptible to theft or diversion. Plutonium is found in the spent fuel of light-water reactors mixed with other fission products which generate much more penetrating radioactivity than does plutonium. As long as plutonium is combined with these irradiated products, their radioactivity inhibits handling and makes it virtually certain that the plutonium will not be stolen.

The threat of plutonium misuse, and the problems of safeguarding plutonium, will increase sharply if any of three events comes to pass:

The plutonium in the spent fuel of light-water reactors is recycled to produce a mixed oxide uranium/plutonium fuel for use in those reactors.

Breeder reactors, which use plutonium as fuel but produce more than they use, are developed.

Nuclear reactors are exported to a large number of countries around the world.

The prospects for development of breeder reactors have been discussed in Chapter 6. The prospects for plutonium recycling and of exports of nuclear reactors can be summarized as follows:

Plutonium Recycling

At this time, most of the plutonium in the spent fuel of nuclear power plants is being stored in fuel assemblies with the other irradiated materials. No reprocessing of these wastes is taking place on a commercial basis. However, industry support for development of a plutonium recycling industry is growing. If that industry does develop, much more separated plutonium will be in circulation, and safeguarding problems will increase. The closed nuclear fuel cycle, including reprocessing, is shown in Figure 7.

Industry and government officials have long argued in favor of recycling nuclear plant wastes to extract uranium and plutonium for later use as a reactor fuel. Recycling, they say, is an important part of the nuclear fuel cycle. It will prolong the supply of scarce uranium resources, lower nuclear fuel costs, facilitate storage and disposal of wastes, and provide eventual fuel for the breeder reactor. The Atomic Industrial Forum goes even further: "The earlier reprocessing and recycle are accepted as integral steps of the nuclear fuel cycle, the earlier nuclear power will gain widespread public acceptance."[5]

A recent report by the Nuclear Regulatory Commission stated that a failure to recycle plutonium could cost the country $18 billion by the year 2000.[6] Industry representatives say the cost would be much greater. The Atomic Industrial Forum predicts that if neither plutonium nor uranium is recycled, consumers of electricity from light-water reactors will have to pay an additional $2 billion a year. The cumulative cost without plutonium, the forum states, will be $50 billion by the year 2000, and will rise to $60 billion if neither is recycled.[7]

Perhaps more importantly, supporters say recycling plutonium will conserve scarce uranium and fossil fuel resources. The Edison Electric Institute told the AEC that the introduction of plutonium recycle could provide a savings "in the order of 10 billion barrels of oil by 1995."[8] Some industry officials say that it could reduce the demand for uranium by 13 to 14 percent. There is serious concern about the scarcity of domestic uranium reserves. Exxon's Ray Dickeman says, "It is probably too late to avoid an important uranium import program."[9] Theodore Taylor, a designer of fission bombs and now a consultant on safeguards, says that without recycling of plutonium or development of the breeder, the trillion-dollar nuclear fission industry will run out of commercial uranium and grind to a halt within two or three decades.[10]

An immediate problem for some utilities is what to do with spent fuel now on hand. Nuclear power plants are required to have facilities for temporary storage of spent fuel, but in general these facilities are limited. The AEC noted in 1974 that ten reactors were storing spent fuel in areas designed for storage of the fuel core in case of an emergency.[11] Several utilities have taken action against General Electric for failing to provide spent fuel storage after the company abandoned its plans for a reprocessing plant in Illinois.

The cause of the storage problem is the uncertainty surrounding the prospects of a commercial reprocessing industry. Originally the AEC and industry had expected that by the mid-1970s two or three irradiated fuel reprocessing plants would be available to separate plutonium and uranium for

FIGURE 7

The Nuclear Fuel Cycle

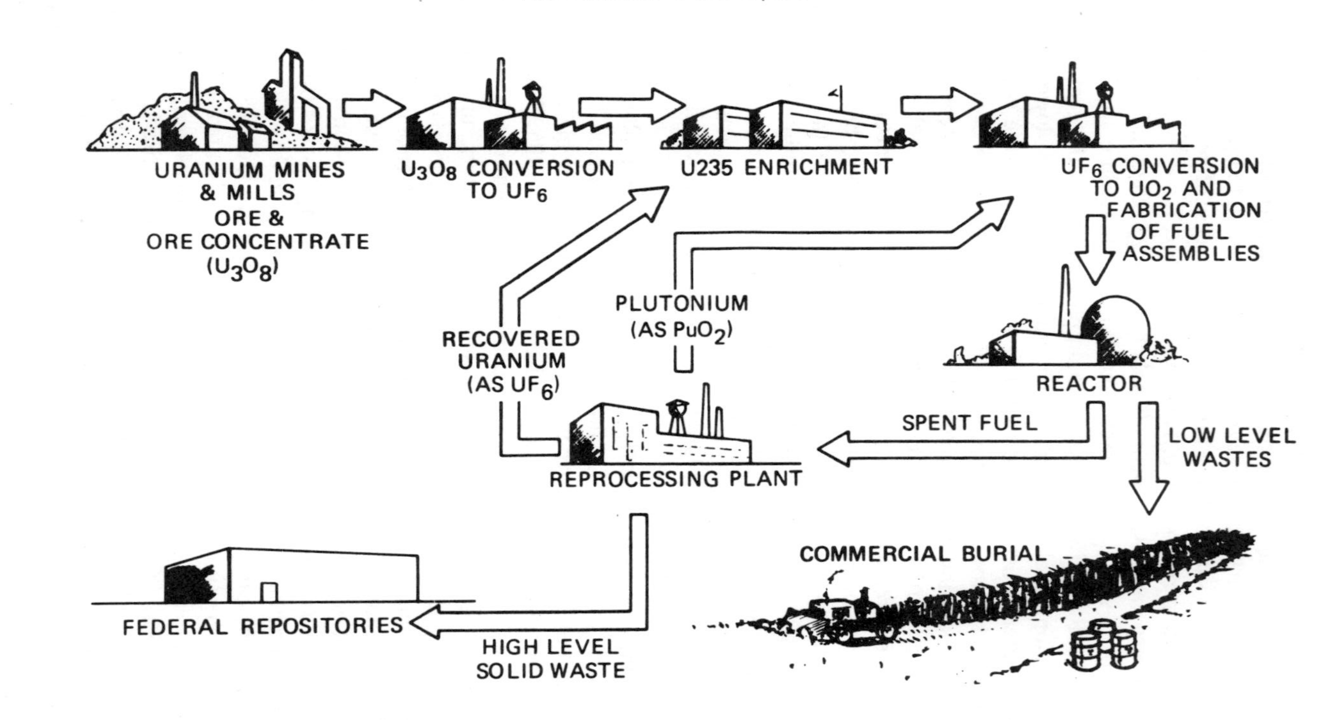

Source: U.S. Nuclear Regulatory Commission, *The Nuclear Industry 1973,* Wash 1173–73.

recycling. However, technical and regulatory problems have contributed to economic difficulties, and the future for private involvement in reprocessing is very clouded. General Electric has abandoned its $75 million reprocessing plant in Illinois because of technical problems. New safety standards which would have forced an investment of $800 million have led Nuclear Fuel Services, a Getty Oil subsidiary, to scrap its West Valley, N.Y. plant. At one time the plant had been scheduled to begin operation in 1975. A group led by Allied Chemical has invested $250 million in a plant at Barnwell, S.C. which was scheduled to begin operations in mid-1976. New regulatory requirements may cost the company an additional $750 million, and there is serious doubt that the plant will ever go into operation without government assistance.[12]

Industry is greatly concerned about the lack of reprocessing capacity in the United States. The Atomic Industrial Forum says the United States needs reprocessing plants in order to "secure nuclear fuel supplies for the U.S. and for appropriate foreign markets." The AIF argues that U.S. and foreign demand will require development of a recycling capability by the early to mid-1980s and estimates that by the year 2000 there should be some nine to ten fuel reprocessing plants, each capable of handling 1,500 to 2,000 metric tons of spent fuel a year.[13] In addition, according to the forum, the United States will need to build shipping casks, transport systems, spent fuel storage facilities, and interim on-site waste storage and waste treatment packaging facilities. It says the total cost will be $15 billion, and adds that this level of investment will require government incentives.

But opposition is growing to government involvement in—or even approval of—reprocessing. Some studies have begun to challenge both the economics of, and the need for reprocessing. A study in 1976 by Pan Heuristics, a division of Science Applications, Inc. in California, states that "the estimated costs to separate plutonium from electric power reactors have increased tenfold in 10 years." Future costs remain highly uncertain, it says. The study estimates that processing plutonium and recycling it into a fuel for light-water reactors will cost 50 percent more than it would to buy fresh uranium. It concludes that "recycling, in short, seems likely to lose rather than save money."[14]

Recycling may only make a minor contribution to fuel conservation. An estimate by the Organization for Economic Cooperation and Development says that the cumulative savings from recycling plutonium to the year 2000 may only be equal to nine months' uranium supply at that date. Pan Heuristics states that based on projections by the Edison Electric Institute, it estimates that fuel savings from uranium and plutonium recycle will amount to about 2 percent of total energy consumption between the years 1975 and 2000.

In light of these studies and the growing concern about the misuse of recycled plutonium, many persons—including President Carter and President Ford before him—are calling for a reassessment of the costs and benefits of recycling and reprocessing. President Ford ordered the Energy Research and Development Administration to conduct an evaluation of the security and economic aspects of reprocessing and recycling; the Nuclear Regulatory Commission issued its final Generic Environmental Statement on recycling plutonium in 1976 and scheduled hearings on the statement in 1977; President Carter has called for a moratorium on sales of reprocessing plants abroad. President Ford said during his term in office that reprocessing is no longer inevitable, and one Ford administration official said of plutonium use in the near future: "Personally, I think it will turn out like oil shale—profitability always just around the corner, but you never quite get there."[15]

Exports of Reactors

Eighteen countries besides the United States now have operable nuclear reactors; another 28 have nuclear plants planned or under construction.[16] By 1980, it is projected, 30 countries will have approximately 230 nuclear reactors, producing around 65,000 kilograms of plutonium a year. By the year 2000, according to some estimates, more than a million kilograms of plutonium could be flowing each year to and from between 2,000 and 2,500 nuclear power plants in some 50 countries.[17]

The boom in foreign countries' interest in nuclear reactors dates from OPEC's decision to raise oil prices. The decision forced a number of countries to recognize their dependence on foreign oil, and it increased their interest in nuclear power as a means to achieving some flexibility of supply. At the same time, oil price inflation and a growing recession were disrupting demand patterns in the United States and other industrial countries. Reactor vendors began to seek new markets, particularly among the developing countries. Projections showed a radical increase in sales to developing countries—and with the increased numbers of reactors, an increase in the amount of plutonium that could be produced. By 1990, one study showed, developing countries could be producing 30,000 pounds of plutonium a year—enough for 3,000 bombs.

Concerns about a proliferation of plutonium production were aggravated by a decision on the part of the United States to cease exports of enriched uranium. In 1974, anticipating a future shortfall of enriched uranium, the United States ended the signing of new export contracts. Thus, countries interested in purchasing reactors have had a greater interest in acquiring reprocessing or enriching technology as well in order to assure themselves of an adequate fuel supply in the future.

The prospect of a rapid proliferation in nuclear technology has focused attention on the U.S. policy governing the export of nuclear plants and materials. Since the launching of its Atoms for Peace program in the mid-1950s, the

United States has been an active proponent of international trade in nuclear technology for civilian uses. The aim was to demonstrate that the atom could be harnessed for peaceful purposes. Through $2.5 billion in loans and guarantees from the Export-Import Bank, the United States has subsidized the sale of reactors to buyers in 11 countries. [18] It trained 1,478 scientists from 41 countries in nuclear engineering between 1970 and 1975, and seven of them were trained in plutonium technology.[19] Both industry and government in the United States have developed a stake in nuclear trade. ERDA has estimated cumulative U.S. export revenues from the sale of nuclear materials at $1.5 billion through 1974. "By 1985," ERDA says, "annual U.S. nuclear power export revenues are projected to range between $3 and 4 billion, increasing by the year 2000 to between $8 and 10 billion. Through the turn of the century, projected cumulative U.S. nuclear power exports are estimated to be between $120 billion and $140 billion."[20] Foreign sales provided an estimated 2 percent of GE's total 1974 revenues (although a negligible percentage of earnings) and 12 percent of Westinghouse's total revenues.[21]

But in the last year, concern about U.S. export policy has been growing. Congress, the administration, and a variety of independent groups and organizations have begun to debate actively the questions of what controls should be placed on foreign sales of nuclear materials. More than a dozen bills relating to proliferation were introduced in 1976, and, for the first time in NRC's short history, one NRC commissioner, Victor Gilinsky, dissented from a commission regulatory opinion—one that granted an export license for sale of a nuclear reactor to Spain. The agreement with Spain, Gilinsky stated, "contains a vital flaw involving the controls over the plutonium—a nuclear explosive—which will be produced in the operation of the reactor." He criticized past U.S. agreements with Spain and commented, "I think it's just unfortunate that we didn't think through some of these problems [control of plutonium] earlier."[22]

There is evidence of support for Gilinsky's position in both the Ford and Carter administrations. Both President Ford and President Carter have urged a tightening of U.S. policy governing the export of nuclear materials, and both have called for a moratorium on the export of reprocessing and enrichment technology. "Given the choice between economic benefits and progress toward our non-proliferation goals," said President Ford, "we have given, and will continue to give, priority to non-proliferation."[23]

Concerns about Exports and a Plutonium Economy

At the heart of all concerns about the proliferation of nuclear power plants, both domestic and foreign, is the fear that plutonium extracted from the spent fuel will be misused. "Unfortunately," President Ford said in 1976, "and this is the root of the problem, the same plutonium produced in nuclear plants can, when chemically separated, also be used to make explosives."[24] The Natural Resources Defense Council foresees a plutonium economy with "perhaps a score of fuel reprocessing and fabricating plants, and thousands of interstate and international shipments containing hundreds of tons of plutonium."[25] The development of a plutonium economy, say both critics and some nuclear supporters, will greatly increase the likelihood that terrorists or foreign governments will obtain the materials necessary to construct a nuclear bomb or a radiological weapon. This may lead to attempts by terrorists, criminals, or irresponsible governments to engage in nuclear blackmail, they fear, and may provide new targets for sabotage efforts. To control these risks effectively, critics say, would be impossible. Too many people and governments will have access to nuclear materials, and efforts to check and monitor all those who have any access will impose a burden costly not only to the public coffers but to citizens' right to privacy.

Concerns over the potential threats can be described as follows:

Nuclear Bomb

Terrorist groups or foreign governments, with the necessary knowledge, could build a nuclear bomb. Terrorist groups, the argument runs, could steal a relatively small quantity of special nuclear materials—10 kilograms of plutonium or uranium 233 or 25 kilograms of highly enriched uranium—and build a crude, transportable nuclear fission bomb. Proponents of this theory believe such a bomb could have an explosive force of 20 kilotons—equal to the Nagasaki bomb, and enough to kill tens of thousands of people and destroy property worth millions of dollars. "The acquisition of special nuclear materials," a special AEC task force on safeguards reported in April 1974, "remains the only substantial problem facing groups which desire to have such weapons."[26]

The AEC task force commented that lack of knowledge was not a problem inhibiting the development of nuclear weapons. "Widespread and increasing dissemination of precise and accurate instructions on how to make a nuclear weapon in your basement, . . . a slow but continuing movement of personnel into and out of the areas of weapons design and manufacturing [creating] larger and larger numbers of people with experience in processing special nuclear materials," have increased the possibility that terrorist groups "are likely to have available to them the sort of technical knowledge needed to use the now widely disseminated instructions for processing fissile materials and for building a nuclear weapon."[27]

With the necessary knowledge, other experts argue, the task of building a bomb is not particularly difficult. Theodore Taylor wrote with Mason Willrich: "Under conceivable circumstances, a few persons, possibly even one person working alone, who possessed about 10 kilograms of plutonium oxide and a substantial amount of chemical explosive, could within several weeks design and build a crude fission bomb. . . . This could be done using materials and equipment that could be purchased at a hardware store and from commercial suppliers of scientific equipment for student laboratories."[28]

Most observers now agree that the greater the number of countries with nuclear reactors, the larger the likelihood that nations will divert plutonium produced by the reactors to manufacture a weapon. The difficulty of reprocessing spent fuel to separate the plutonium from other irradiated elements acts as a temporary restraint on plutonium use for weapons development. For a number of years some observers hoped that this restraint, together with international safeguards agreements, would provide an effective disincentive to diversion. India's nuclear explosion in 1974, using plutonium produced by a reactor given by Canada and fueled with "heavy water" from the United States, destroyed that hope.

In fact, many persons consider it relatively easy for countries to develop a reprocessing capacity. Experts estimate that Japan, India, Canada, Taiwan, Argentina, Spain, the Euratom countries (West Germany, Belgium, Italy, and the Netherlands) and Israel—as well as the five original nuclear powers (United States, Britain, Soviet Union, France, and China)—already have reprocessing capabilities. And, in the heat of competing for the nuclear market, several countries have shown a willingness to include reprocessing technology as a "sweetener" in a package sale of reactors. France included a reprocessing plant in its sales of reactors to Pakistan, Iran, and South Korea—although U.S. pressure has succeeded in getting South Korea to remove reprocessing from its agreement—and, following U.S. protests, reprocessing may be excluded from the sales to Iran and Pakistan.

Whether reprocessing is included in the sales package may not make much difference in the long run, however. Small-scale reprocessing plants can be built for $10 to 50 million,[29] and the key factor affecting their development is more likely to be international pressure than either technical or capital constraints. When U.S. intelligence reports indicated that Taiwan had been secretly reprocessing uranium, the State Department stated that continuing efforts to develop a reprocessing capacity would "fundamentally jeopardize" U.S. nuclear cooperation with Taiwan. Taiwan agreed to halt all activities relating to reprocessing.[30]

Less threatening than diversion of nuclear materials by a national government, but still of concern to many critics of the plutonium economy, is the possibility that terrorist groups will steal enough plutonium to construct a bomb. Brian Jenkins, manager of the Rand Corporation's study project on terrorism, says the prospect of terrorists using nuclear weapons cannot be ruled out, but making an accurate assessment of the probability that a specific kind of event will occur is impossible. It is important, he says, to put terrorists' objectives into proper perspective. Terrorists generally have made limited efforts to seek support, to attract public attention, or to use fear in order to obtain specific objectives. They have had access to materials—poisons or conventional explosives—which in certain situations could have produced large numbers of casualties. Thus far, he notes, apparently they have concluded that their objectives would not be advanced if they used these potent weapons to create incidents of mass destruction.

But Jenkins says it is possible that, under certain circumstances, the level of violence employed in acts of terrorism will change. If terrorists think that governments and populations are growing inured to lower levels of violence, he says, they may decide that more dramatic actions are required to attract attention to their cause. It also is possible, he says, that because of inbred fears of atomic devices, nuclear weapons would prove more attractive to terrorists than other esoteric forms of destruction. For example, he suggests, a theft of plutonium is likely to receive more publicity than a theft of chemical poisons, and the threat of an atomic bomb would cause greater concern to the general population than a threat of a fire in a petroleum tank farm.[31]

The successful construction of a bomb, either by terrorist groups or by an irresponsible national government, poses an immensely serious threat to world peace, nuclear opponents say. They are skeptical that adequate safeguards against the development of such a bomb can be developed. In any event, they say, the risk created is unacceptably large.

Radiological Weapons

Smaller quantities of plutonium could be used to build dispersal weapons that would lethally contaminate very large volumes of air. Plutonium emits alpha particles that have very little penetrating power in comparison with gamma or X rays, and hence it can be handled with minimal shielding for hours without hazard. If taken into the bloodstream, however, plutonium becomes one of the most carcinogenic agents to be found. It is particularly dangerous if inhaled. Particles weighing some 10 millionths of a gram, if inhaled, can cause death from fibrosis of the lungs within a few weeks.[32] According to Ralph Nader, "an amount of the lethal plutonium-239 not exceeding 20 pounds, if efficiently dispersed, could give lung cancer to everyone on earth."

The possibility of dispersal this efficient is surely remote. Nevertheless, Nader's example serves to emphasize that very small quantities of plutonium can cause tremendous harm. It is estimated that even a few grams dispersed through the air-conditioning system in an aerosol suspension could pose a serious threat to the occupants of a large office building or enclosed industrial facility. Effective dispersal in open areas would be more difficult, but experts calculate that a few dozen grams of plutonium could contaminate several square kilometers sufficiently to require the evacuation of people in the area and necessitate a very difficult and expensive decontamination operation.[33]

Nuclear Blackmail

Some experts believe that an expanding nuclear industry will make the theft and misuse of nuclear materials seem so plausible that criminal groups might be able to extract huge ransoms from threats of nuclear explosion or radiation without actually having possession of any lethal materials or without constructing a bomb. An incident that occurred in Orlando, Fla., in 1970 is frequently cited in support of this view. The mayor of Orlando received a note threatening destruction of the city by a nuclear bomb and demading $1 million and safe passage to Cuba. Accompanying the note was a diagram of a crude bomb which some experts thought should be taken seriously. Both federal and municipal authorities took the threat seriously, until a local high school student explained it was a hoax he had perpetrated. Taylor argues that the Orlando incident demonstrates not only that a credible threat can occur, but that a bomb was not necessary in order for the threat to be effective.[34] The potential effectiveness of such a threat may encourage a black market in plutonium. Taylor and Willrich estimate that commercial value of nuclear weapons materials will be between $3,000 and $15,000 per kilogram—roughly equal to the price of heroin on the black market. "The same material might be hundreds of times more valuable to some group wanting a powerful means of obstruction," or as a means of seeking ransom.[35]

Nuclear Sabotage

Several observers also believe that the proliferation of nuclear power plants and fuel reprocessing and fabrication facilities will greatly increase the likelihood that attempted acts of sabotage will occur, with potentially catastrophic effects. Some fear that trained saboteurs using explosive devices could cause the dispersal of highly irradiated nuclear fuel that is in transit or in facilities where spent fuel assemblies are stored or reprocessed. Others are concerned that a skyjacked aircraft might be used to crash into a nuclear reactor, causing a core meltdown and consequent release of intense radioactivity. They say power plants are designed to sustain the impact of a 200,000-pound aircraft, and a Boeing 747, for example, weighs 365,000 pounds. Still others see a danger in the possible takeover of a nuclear power plant by a band of highly trained, sophisticated terrorists who might destroy it in such a way as to kill thousands, perhaps millions, of people. A General Accounting Office report concluded in 1964 that current security measures at nuclear power plants could not prevent a takeover by as few as two or three armed individuals.[36] Rand's Brian Jenkins cites nine cases in which acts or threats of sabotage involving radioactive materials have occurred in the last five years. "The rapid growth of the nuclear industry," he concludes, "increasing traffic in plutonium, enriched uranium, and radioactive waste material, the spread of nuclear technology both in the United States and abroad—all increase the opportunities for terrorists to engage in some type of nuclear action."[37]

THE SAFEGUARDS CONTROVERSY

No one denies the need for safeguards against the misuse of plutonium, and most observers admit that current safeguards are inadequate. President Ford, for example, stated in October 1976 that "the standards we apply in judging most domestic and international [nuclear] activities are not sufficiently rigorous to deal with this extraordinarily complex problem."[38] But there are points of serious disagreement on how great the threat of misuse of plutonium is; whether effective international safeguards can be established; whether adequate domestic safeguards can be instituted; how expensive effective safeguards will be and who will cover the costs; and whether safeguards measures will endanger individuals' civil liberties.

Proponents and opponents of nuclear power agree that safeguards questions present the industry with particular and important problems, but opponents in general feel that nuclear power should be slowed down—or even brought to a halt—until the problems are resolved, and supporters generally are more confident that institutions, systems, and technology can meet the concerns that have been raised. Many critics assert that proponents do not comprehend the dimensions of the problem; proponents say that critics have exaggerated the problem. Almost universally, proponents of nuclear power maintain that it would be a mistake to slow the momentum of nuclear power development because of safeguards problems. On specific questions, these are the stands of both sides:

Threat of Misuse

The strongest critics of nuclear power argue that the dangers of plutonium misuse are so severe that further development of nuclear power should be abandoned. If it is not, they say, pressure from utilities and manufacturers almost certainly will lead to development of a nuclear fuel recycling industry and increased exports, and it may lead to development of breeder reactors (if the technical and economic problems in the breeder program can be solved). Moreover, if the United States continues to pursue domestic expansion of nuclear power, it will not be able to sponsor with credibility any slowing of development of nuclear power abroad.

Some persons concerned with safeguards focus their attention on stopping recycling, the breeder reactors, or exports of nuclear power plants. Others do not oppose development of a domestic or an international plutonium economy but argue that much more attention must be devoted to strengthening safeguards if such development is to proceed.

The concern that national governments may decide to divert plutonium from a civilian reactor for use in weapons manufacture has grown dramatically in the last two years—since the explosion of a nuclear device by India in 1974. The rapid expansion of the export market, and particularly the potential market in developing countries, has begun to trouble many people, even those supporting nuclear power. David Lilienthal, the first chairman of the Atomic Energy Commission, has commented on the "impending disaster" of the spread of nuclear weapons capabilities through sales of nuclear facilities abroad. "Proliferation of capabilities to produce nuclear weapons of mass destruction is reaching terrifying proportions," Lilienthal testified in January 1976. He recommended a temporary ban on the export of nuclear devices and materials.[39]

Critics of the export program say that one aspect of the problem is the difficulty of assuring long-term security. A recent study by Pan Heuristics states that "the most immediate prospects for acquiring nuclear explosives tend to be small or less-developed countries, especially those outside the Soviet orbit, not—as once expected—the most advanced industrial powers." The authors argue that "some of the countries that may soon acquire nuclear weapons . . . are politically unstable and much more liable to sudden threats of mass destruction from dissident factions." South Korea, Taiwan, Pakistan, Iran, Brazil, Argentina, Spain, Yugoslavia, and Rumania are among those cited as close to a decision to develop a nuclear weapons capability. They are also, the study says, countries that appear to lack the security of a long-term alliance with a nuclear power or whose permanent position in such an alliance is subject to some doubt.[40]

Even if a country is stable at the moment, in a secure relationship, and is apparently able to offer assurance of adequate safeguards, it is possible that the situation may change in the future. As the number of countries with nuclear plants and a capability for reprocessing the spent fuel increases, so does the likelihood that one of these countries will decide to divert plutonium from the spent fuel to develop a weapon. Former AEC Commissioner Henry Smyth comments that a stockpile of plutonium may build up under a stable government. "Now suppose there's a revolution. A totally new and crazy government comes in, and there's the plutonium just sitting there asking to be made into a bomb."[41] In describing the problem, Hans Bethe quotes Enrico Fermi, "the father of the atomic age: 'every country has a mad dictator every few centuries.' "[42]

In general, industry and government officials agree that the threat of diversion by foreign governments in order to make nuclear weapons is real and serious, and they support efforts to tighten international safeguards to prevent such misuse. But they are not as concerned, as are many critics, about the possibility that terrorists will succeed in stealing enough material to build a bomb. Government officials, for instance, disagree with the statements by Taylor and others about the ease with which a nuclear bomb can be constructed. They say it would take a half-dozen or dozen "quite competent, very skilled scientists, engineers, electronic people to fabricate such a device." The need for extremely tight security during the entire undertaking, they say, would further reduce the chances for a successful outcome. They do not rule out the possibility that a theft might occur, but they consider it less likely than does Taylor.[43]

Government and industry officials also argue that the nuclear industry has been singled out unfairly by critics as a potential target for sabotage or theft. There are other, more attractive targets, they say, involving materials that are easier to obtain than plutonium. For example, one industry representative said, nonnuclear bombs could be placed in large population centers; germs or LSD could be used to contaminate water supply systems; or an oil tank farm, natural gas storage area, or ship filled with propane gas could be attacked with explosives. The efforts being made to protect against thefts of nuclear materials, proponents say, are more extensive than the security measures being taken against more conventional manmade disasters, and the risk of blackmail or sabotage from nuclear materials should be no greater.

International Safeguards

Those who oppose export of nuclear materials do so for three specific reasons: (1) foreign governments, particularly in countries with weak law-enforcement capabilities, extreme political ideologies, or active terrorist organizations, may not be capable of protecting nuclear materials; (2) the government of a country with a civilian nuclear power program may choose to divert some of the materials to develop nuclear weapons, as India did in 1974; and, perhaps most important, (3) the International Atomic Energy Agency (IAEA), the multicountry agency working to monitor the use of nuclear materials, is not capable of regulating international nuclear activity.

As mentioned earlier, when the United States launched its Atoms for Peace program in the mid-1950s, the hope was that the sharing of nuclear power for peaceful purposes would help to mitigate against its being used for war. Access to the new technology, which was then virtually a U.S. and Soviet monopoly, would be tied to international controls that would protect against its misuse.

In 1958 the International Atomic Energy Agency was established to carry out the inspection of nuclear facilities. Its role was broadened a decade later by the negotiation of the treaty on the nonproliferation of nuclear weapons. Under the terms of the treaty, the 97 signers, in states without nuclear weapons, agreed to allow the IAEA to monitor all nuclear activities. In exchange, it was understood that they would receive preferential access to nuclear power —including the technology for enriching uranium and for recycling plutonium. As Leonard Ross of the California Public Utilities Commission wrote recently in The New York *Times Magazine,* "The faith in safeguards had not yet been shaken."[44]

The Indian explosion seriously shook that faith. Observers pointed out that there are major loopholes in the nonproliferation treaty. The signer has the right to conduct any nuclear activity short of producing a nuclear weapon, and any signer can withdraw from the treaty after giving 90 days' notice. A country may go a long way toward development of a weapons capacity but stop just short of the final step which would result in violation of the nonproliferation treaty. It is conceivable that the country could then withdraw from the pact and proceed with the last actions required to manufacture a weapon. The problem is accentuated by what some experts call the "shrinking critical time" required to make a nuclear explosive. "The fundamental problem," Albert Wohlstetter wrote in a Pan Heuristics report on proliferation, "is that, for an increasing number of 'non-weapons' states, the critical time to make an explosive has been diminishing and will continue to diminish without any necessary violation of the clearly agreed-on rules—without any 'diversion' needed—and therefore without any prospect of being curbed by safeguards that have been elaborated for the purpose of verifying whether the mutually agreed-on rules have or have not been broken."[45]

There is also a problem with the enforcement capacity of the International Atomic Energy Agency. The agency does not have the capacity to enforce agreed-upon international safeguards. There are 40 inspectors for 112 reactors operating in 20 countries.[46] Nor does it have the authority. According to Senator Adlai E. Stevenson, III (D-Ill.), the IAEA provides "little more than an inventory accounting system" which can detect diversions after or as they occur but is "powerless to prevent them from happening." [47] It is, according to Fred C. Ikle, director of the Arms Control and Disarmament Agency, "a burglar alarm, but not a lock."[48] And the IAEA has no sanctions authority to call into play in the event that the alarm is sounded.

In response to these concerns, Stevenson has called for a one-year embargo on the sales of nuclear reactors, Secretary of State Henry Kissinger proposed a system of multinational control over reprocessing plants, and David Lilienthal argued for a temporary embargo on the export of nuclear devices and materials. Lilienthal's proposal has won the approval of a number of proponents of nuclear development, including Hans Bethe. "When I first heard about it and read it, I didn't like it," Bethe stated in early 1976, "but now I like it when Mr. Lilienthal said the embargo was temporary until we worked out real controls. But we have to make clear that the embargo is temporary until a treaty can be concluded between nuclear countries that really assures control over proliferation."[49] Concern over proliferation controls has led to a series of meetings between major nations exporting nuclear materials. A U.S. decision to halt unilaterally the sale of all reprocessing and enrichment technology has been followed by France's decision to reconsider the sales of reprocessing plants to Iran and Pakistan.

Until recently, many proponents had considered an embargo to be unnecessary. Some argued that the IAEA safeguards would be sufficient. Others were fatalistic—if the United States did not sell the technology, other countries would, and without safeguards as strict as those posed by the United States. Since the U.S. and French decisions, agreement seems to be widening that, at least for the time being, a halt to the export of sophisticated nuclear technology is justified.

In order to stop such exports, and still be able to maintain a position in the export market for nuclear reactors, industry officials argue that it will be necessary for the United States to expand its own enriching and reprocessing facilities. The United States must be able to provide "attractive alternatives" to countries that might otherwise be interested in their own reprocessing or enriching. "If supplier nations make available an assured supply of enriched uranium as well as fuel reprocessing and waste disposal services at competitive prices, new fuel cycle ventures will appear less necessary," according to the Atomic Industrial Forum.[50] Whether the United States should develop a domestic reprocessing capability and what methods it should use for disposal of waste remain questions of great debate. One aspect of the debate is whether domestic safeguards are adequate if a reprocessing industry develops.

Domestic Safeguards

Critics on this issue agree with the conclusion in 1973 of an AEC safeguard study group that existing safeguard systems are "entirely inadequate to meet the threat," and that developing adequate safeguards will be a most difficult task.[51] Current safeguards systems, comments Theodore Taylor, "simply do not go anywhere far enough to prevent thefts by groups of people with at least the skills and resources that have been used for major thefts in the past."[52] The AEC study group concluded that they were not adequate to protect against the threat that would be posed by "15 highly trained men, no more than three of which work within the facility or transportation company from which the material is to be taken."[53]

To develop adequate safeguards, says Taylor, "is among the greatest challenges humanity ever faced." He cites the dangers of the development of a black market for weapons-grade materials.[54] Clarence Larson, a former AEC commissioner, suggests: "Once special nuclear material is successfully stolen in small and possible economically acceptable quantities, a supply-stimulated market for such illicit material is bound to develop. And such a market can surely be expected to grow, once the source of supply has been identified."[55]

One major problem will be to develop technical devices that will allow accurate and continuous accounting of plutonium inventories. The director of the AEC's office of safeguards materials management once commented, "We have a long way to go to get into that happy land where one can measure scrap effluents, products, inputs and discards to a 1-percent accuracy."[56] In the past, critics point out, large amounts of radioactive materials have "disappeared" from inventories and have not been found. The General Accounting Office (GAO) reported in July 1976 that tens of tons of nuclear material, much of it weapons-grade, was missing in 34 uranium and plutonium processing plants around the country. In commenting on the GAO report, members of the staff of a House subcommittee said they seriously questioned whether the accountability procedures established by ERDA can give "reliable indications of whether bomb quantities of special nuclear materials have been lost or stolen."[57]

Many NRC and industry officials agree that safeguards have not always received the attention they deserve, but most contend that times have changed. An ERDA official, James Liverman, has said that "we really only started upgrading our security in 1972, after the terrorist attacks on the Munich Olympic games."[58] In the beginning, according to one Atomic Industrial Forum official, keeping track of nuclear materials was viewed as an accounting problem. Then-AEC Chairman Dixy Lee Ray told a Senate subcommittee in 1974 that "the commission has recently acted to improve the safeguards, and we will continue to modify them to meet changing levels of threat."[59] L. Manning Muntzing, the NRC's director of regulation, has commented, "We think that, properly implemented, the regulations which we now have in place are appropriate for today's needs."[60]

Romano Salvatori of Westinghouse says that "regulations currently being implemented by the nuclear industry provide more than adequate assurance that this material will not be readily susceptible to theft or diversion . . . The material accountability requirements provide protection against the undetected diversion of special nuclear materials. Protection against actual theft is provided by physical protection systems. . . . The risks of theft of special nuclear materials have been identified, and appropriate accounting and security systems have been developed to reduce these risks."[61]

Nevertheless, many critics argue, any conceivable safeguards system will suffer from the same basic weakness—the fallibility of man. Alvin Weinberg —a supporter of nuclear power—has noted that a "continuing tradition of meticulous attention to detail" will be required to protect nuclear wastes. He suggests that the establishment of a "scientific priesthood" might be required.[62] Others argue that no "priesthood" made up of as many people as might have access to special nuclear materials can be trusted completely. For example, the Natural Resources Defense Council writes, "the New York Police Department has proven incapable of maintaining security over convis-cated heroin. Are similar losses of plutonium acceptable?" The Council argues that it will be difficult for any institution charged with constant monitoring to maintain an acceptably high level of vigilance. It fears that with time, the institution will become complacent and safeguards will loosen. "Our experience indicates that rather than sustaining a high degree of esprit, vigilance and meticulous attention to detail our governmental bureaucracies instead become careless, rigid, defensive and, less frequently, corrupt."[63]

Costs of Safeguards

Another problem, critics say, may be the costs of safeguards technology and the manpower required to form federal or private security forces. Although they make no estimates of the costs of a safeguards program which they would consider adequate, critics say the costs could be significant. Safeguards, like safety research and storage, have been subsidized, a spokesman for National Intervenors comments. The subsidy must be counted as a cost, he argues, and it may affect the competitiveness of the nuclear option.

Industry officials also are concerned over how the increasing costs of safeguards programs will be allocated. A statement issued in 1974 by General Atomic (whose high-temperature, gas-cooled reactors will use highly enriched uranium as fuel) argues that the "cost of maintaining a federal protective service should be borne by the taxpayer as a special type of police arrangement which the national security interest requires. The security needed is clearly in excess of that required to protect the property interests of the owner of the material and, in fact, is directed at protecting the security interests of the body politic." Moreover, General Atomic says, a federal storage facility should be established to ensure that "private industry is not given the burden of storing quantities of special nuclear material in excess of what is required for reasonable working inventory."[64]

The other alternative for allocating costs would be to pass them on to the consumer in the form of higher utility rates. Just how much these increased costs would amount to is difficult to predict. E. R. Johnson estimates that the total cost of safeguarding plutonium would amount to $2,450 per kilogram in 1980, although it might drop to $1,625 per kilogram by 1990. He says that safeguards would represent less than 4 percent of the cost of reprocessing plutonium, and less than 1 percent of the total per-kilowatt cost of generating power from a nuclear plant.[65]

In any event, proponents maintain that safeguards costs will not seriously affect the competitiveness of nuclear power. Taylor and Willrich argue that even if total costs for safeguards were to rise to $800 million a year for the entire industry, which they believe is most unlikely, "this would hardly be sufficient by itself to affect substantially the over-all economics of nuclear power."[66] An official of the Atomic Industrial Forum has asserted that it is "possible to protect special nuclear material sufficiently so that reasonable people agree that any risk from diversion or sabotage is negligible . . . at a cost which, although high, need not be so high as to cripple the economics of nuclear power."[67]

One way to reduce safeguards costs would be to combine some or all of the facilities involved in the nuclear fuel cycle—fuel fabrication, reprocessing, and power reactors—in "nuclear parks." These would eliminate some vulnerable transit stages and enable a consolidation of physical protection requirements. Taylor estimates colocation of fuel reprocessing and fuel fabrication facilities could lower safeguards costs by about 10 percent. Location of all nuclear facilities in a nuclear park would result in an even greater cost reduction, although environmentalists suggest that nuclear parks may have severe environmental costs which would require large quantities of water for cooling.

Civil Liberties

Some critics believe that if a safeguards system were to be implemented, it would infringe unacceptably on many persons' civil liberties. The Natural Resources Defense Council maintains: "An adequate safeguards [system] would not be acceptable in practice . . . due to the tremendous cost of such a system in terms of human freedom and privacy." Council representatives argue that implementation of safeguards measures of the kind suggested in the draft impact statements on plutonium recycling—repeal of legislation barring illegal searches and institution of martial law in cases of nuclear blackmail threats—poses a basic question: "Will the plutonium economy raise socio-political problems of such magnitude that their resolution will be unacceptable to society?"[68] In a lengthy treatment of the civil liberties impact of plutonium recycling, in the *Harvard Civil Liberties Law Review,* Russell W. Ayres concludes that "the challenge to the legal system's competence to adjust social interests in public safety with individual interests in civil liberties may be the most significant social cost of plutonium."[69]

Industry spokesmen acknowledge that some of the security measures that would be required, such as mandatory security clearances for people working with special nuclear materials, raise the threat to civil liberties. They argue with Carl Walske of the Atomic Industrial Forum that it is up to industry "to guard vigilantly against any such abuse." They do not expect that any monitoring of citizens that might be required would be excessive, and they see this minimal loss of privacy as a small price to pay for a continued supply of cheap electricity.

DISPOSAL OF RADIOACTIVE WASTES

Critics of nuclear power development also are extremely concerned over provisions for handling the radioactive wastes from nuclear power plants which must be isolated for many thousands of years. The Natural Resources Defense Council comments that "the environmental and health hazards posed by the generation, transportation and eventual disposal of extremely toxic, long-lived radioactive wastes are unparalleled in the history of man."[70]

Nuclear Wastes

Nuclear power plant wastes can be divided into three general categories: high-level wastes, transuranic contaminated wastes, and low-level contaminated wastes. High-level wastes, although smaller in volume than the other two categories, offer the greatest risk and pose the greatest problems of transportation and storage.[71]

High-Level Wastes

Fission products in high-level wastes emit gamma rays which have a highly penetrating radiation and require massive shielding in transit or in storage. The longest-lived of the fission products, cesium 137 and strontium 90, have a half-life of 30 years and become harmless after 800 to 1,000 years.

High-level wastes also include transuranic elements, the most dangerous of which is plutonium. Transuranic elements emit alpha particles which have little penetrating power and, consequently, require little shielding. But some, like plutonium, have a high specific radio-toxicity and an extremely long half-life. Plutonium has a half-life of 24,000 years and is considered to present a potential hazard for at least 250,000 years, and possibly as long as a million years.

Most of the high-level wastes in the United States have been produced by the nation's nuclear weapons program. Since the 1940s, the nuclear weapons program has produced 215 million gallons of high-level nuclear wastes. Of this amount, 80 percent has been solidified, and in January 1976 the military's high-level waste inventory was "about 75 million gallons, half of which was in solid form," according to a report by Mason Willrich.[72]

Commercial reactors have produced only 600,000 gallons of liquid high-level wastes. By the year 2000, the nuclear industry may have generated 60 million gallons of liquid high-level wastes, but Willrich estimates that it will not be until 2020 that "the commercial power industry will have produced as much liquid high-level waste as has been generated by U.S. military programs."[73]

Most experts agree that the problems of handling high-level wastes result not so much from quantity, but from the difficulty in developing a suitable method of long-term isolation and confinement. A 1,000-megawatt power plant will produce only about 60 cubic feet of solidified high-level waste annually.[74] Willrich reports that the volume of all the solidified high-level waste produced commercially through the year 2000 will be equal to a cube 70 feet on a side. That produced by the military will be equal to a cube about 220 feet on a side.[75]

Transuranic and Low-Level Wastes

Transuranic wastes contain many of the same elements as high-level wastes, but in smaller quantities and without the high-level penetration of radiation. Before 1970 no distinction was made between transuranic and low-level wastes (items such as used reactor equipment with filters or protective clothing) which do not emit radioactions but may have been contaminated by radioactive materials, and both were buried. Since then, however, transuranic wastes have been stored above ground in a retrievable fashion which would allow for eventual disposal in a permanent facility. By June 1974, 42 million cubic feet of transuranic and low-level wastes generated by military programs and 9 million cubic feet from commercial activities had been buried in commercial sites. By the year 2000, Willrich writes, the commercial industry may have generated 50 million cubic feet of transuranic wastes.[76]

The Waste Management Program

The responsibility for the development of waste storage facilities has been divided among a number of government agencies, each of which is responsible for a particular aspect of the program.[77]

ERDA has prime responsibility for the development of waste management technology, for construction and management of federal waste depositories.
The NRC will set the safety standards for the storage facilities.
The U.S. geological survey will aid in site selection.
EPA will establish environmental standards.

ERDA declared in 1976 that the only practical solution to long-term storage of high-level waste appeared to be "placement in a stable geological formation at depths reachable by conventional mining methods."[78] The best locations for federal storage facilities, officials say, are in Western salt beds, nearly 3,000 feet under ground, which have remained undisturbed geologically for millions of years. Geological Survey is investigating possible sites. Core drilling at 20 to 25 sites may begin in 1977, and construction of a pilot storage facility may get under way by 1978. ERDA is to begin testing techniques in 1978 for reducing nuclear wastes to a solid form similar to pyrex, and the NRC is to complete work on waste storage and transportation models and to assess storage container options. The test "emplacement" of nuclear fuel is scheduled to begin in the mid-1980s, according to the current schedule.[79]

THE WASTES CONTROVERSY

Persons on all sides of the nuclear power question agree that development of a system of permanent disposal is desirable. Nuclear wastes that are stored in a manner that makes them retrievable also could be misused or might be released accidentally into the environment. The current method of storage—by utilities in ponds—is inadequate. In December 1975 the NRC reported that there were 6,047 fuel assemblies stored at nuclear plants.[80] Not only will delay in development of storage facilities force some utilities to build new storage areas, but the lack of a decision on storage, many proponents say, makes it more difficult to convince the public of the safety of nuclear power. "The most important problem facing nuclear power is waste management," Senator John O. Pastore (D-R.I.) has stated. "Reactors are safe, and the bugaboo will vanish. But the radioactive waste has to be stored for thousands of years."[81]

Views of Nuclear Opponents

Critics of nuclear power argue that until government and industry have developed and tested a technology for safe, permanent containment of radioactive wastes, it is irresponsible to continue producing them. Their concern is heightened by two factors: the extremely long periods during which transuranic wastes remain dangerous and should be safeguarded, and what they view as the unsatisfactory record of the AEC and the NRC in handling wastes. "All in all," the Union of Concerned Scientists has stated, "we are profoundly disturbed by the prospects of accumulating radioactive wastes that the country does not have the technology to cope with and which the AEC is today dealing with in a supremely careless manner."[82] With particular regard to the breeder reactor, critics and many experts feel that industry and government have considerably underestimated the quantity of wastes contaminated by plutonium which would be generated by the breeder-reactor fuel cycle because of the higher incidence of plutonium in the full cycle.

The most persistent concerns regarding nuclear wastes stem from the need to isolate these wastes for extraordinarily long periods of time. Many feel that a span of even 800 years transcends man's ability to predict events. Hannes Alfven says: "The problem is how to keep radioactive waste in storage until it decays after hundreds of thousands of years. The deposit must be absolutely reliable as the quantities of poison are tremendous. It is very difficult to satisfy these requirements for the simple reason that we have had no practical experience with such a long-term project."[83]

Others point out that virtually none of man's political or social institutions has remained stable and free of revolution for as long as 1,000 years. Alfven sees the need for a society "with unprecedented stability." The Center for Science in the Public Interest states that "no society in history has lasted an instant compared to the long times involved in the nuclear waste problem."[84] The Union of Concerned Scientists declares that "institutional arrangements do not exist and never have existed to guarantee the monitoring of or attendance upon storage facilities over a millennium."[85]

Does man have the right, these critics ask, to create a danger of potentially cosmic proportions which will, in effect, last forever? Some feel strongly that the answer is no. The National Council of Churches has condemned the use of plutonium and has said that the problem of waste storage creates a "fundamental ethical question" of one generation's right to burden another with "an element of risk comparable to that of our vast store of nuclear arms."[86] The Center for Science in the Public Interest alleges that a policy of promoting nuclear power "follows the principle of eating, drinking and being merry for the earth has no tomorrow. It takes the chance of compromising the earth as the home for succeeding generations."[87]

In support of their point of view, many critics argue that efforts to deal with the problem of waste disposal have been inadequate. For example, Thomas C. Hollocher, speaking for the Union of Concerned Scientists, contends that "the proposals so far considered seriously by the AEC for the disposal of wastes are dubious in concept, not technically feasible, or they are so dependent upon site-specific geological characteristics that suitability cannot be determined a priori without extensive on-site investigation."[88] Mason

Willrich concluded in 1976 that "the existing organization for radioactive waste management will be unworkable if left unchanged, [and] the existing framework for radioactive waste regulation will be ineffective if left unchanged."[89]

Many critics agree with Willrich, who wrote recently that "thus far the U.S government's record on management has been marred in a sufficient number of instances to be a cause of concern."[90] They point out, as an example, that 18 leaks from tanks at the government facilities in Hanford, Wash., have resulted in the loss of 430,000 gallons of radioactive wastes.[91] Critics view the AEC's initial plans for storing wastes in salt mines in Lyons, Kans., as a further example of the AEC's eagerness to find a solution outdistancing its technical capacity. The AEC was forced to abandon its plans for storage after it discovered adjacent salt mines and old exploratory oil and gas wells which jeopardized the stability of the site's salt formation. If the wastes had been deposited in the salt mine before this had been discovered, recovery and removal of them might have been impossible.

Views of Nuclear Proponents

Supporters of nuclear power development generally argue that, when viewed in context, disposal of high-level wastes does not pose a danger substantially different from any that man already lives with. They cite, as examples, the many hazardous materials (such as poisonous gases) already stored on the earth that are subject to natural disasters of one kind or another and the many man-made structures, such as large dams, that might one day suffer damage causing untold tragedy. Society has accepted such hazards in the belief that the benefits derived far outweigh the risks, they say. So it is with nuclear power; what society gains more than compensates for the risks. Weinberg summed up this attitude when, after describing the risks of nuclear power, he concluded: "What we offer in return, an all but infinite source of relatively cheap and clean energy, seems to me well worth the price."[92]

Those in favor of nuclear power also contend that critics have greatly exaggerated the potential risks of nuclear waste disposal. They argue that a 1,000-megawatt power plant using coal produces 50,000 tons of bottom ash and 250,000 tons of fly ash per year; a similar plant using oil produces 1,400 tons of ash; and an equivalent-size nuclear reactor generates 1.4 to 2.3 tons of solidified high-level wastes. Supporters of nuclear power are confident that finding a method for permanent disposal of such a relatively small quantity presents no insurmountable problem.

Government officials maintain that the methods being used at Hanford are not comparable to those planned for future storage and that the 30 years of experience they have had with radioactive wastes from the nuclear weapons program gives them "total confidence" in their ability to design and construct a safe and secure repository.

A recent ERDA statement says that the record "shows steady progress in reducing waste volumes, releases of radioactivity to the environment, improved storage tanks, and planning for long-term disposition of the waste. These improvements do provide strong support to ERDA's contention that radioactive waste from commercial power reactors can be handled safely. The bank of knowledge is growing and systems have been constantly improved with time."[93]

Industry officials agree. The Atomic Industrial Forum has written that "in many ways the storage and disposal of radioactive wastes are much less a problem than dams and dikes—technically speaking, even less a problem than the management of our cities' garbage and sewage."[94] Carl Walshe, president of the forum, has stated that "the technology for nuclear wastes isn't a problem. The difficult part will be getting over the political hurdle, convincing people that it's not that bad."[95]

NOTES

1. J. Gustave Speth, Arthur R. Tamplin, and Thomas B. Cochran, "Plutonium Recycle: The Fateful Step," *Bulletin of the Atomic Scientists,* November 1974, p. 15.
2. Nuclear Regulatory Commission, "Final Generic Environmental Statement on the Use of Recycle Plutonium in Mixed Oxide Fuel in Light-Water Cooled Reactors," NUREG—0002, Vol. 2, August 1976, p. II–31; Also, Theodore B. Taylor and Mason Willrich, *Nuclear Theft: Risks and Safeguards* (Cambridge, Mass.: Ballinger, 1974), p. 64.
3. Ibid., pp. 60 ff.
4. Ibid.
5. Atomic Industrial Forum, "The Importance of Closing the Nuclear Fuel Cycle," October 1, 1976.
6. NRC, op. cit., Vol. IV, p. XI–1.
7. Atomic Industrial Forum, "General Comments on GESMO," October 28, 1974.
8. W. Donham Crawford, EEI, letter to S. H. Smiley, AEC, October 25, 1974.
9. Quoted in "Fuel Shortage Forecasts for U.S. Nuclear Plants Within a Decade or Two," *Wall Street Journal,* June 7, 1976, p. 1.
10. Interview with IRRC.
11. AEC memorandum, October 22, 1974.
12. Sarah Miller, "Recycling the Nuclear Debate," New York *Times,* November 7, 1976, p. 2; also Jeffrey A. Tannenbaum, "Big Plant to Recycle Nuclear Fuel is Hit by Delays, Cost Rises," *Wall Street Journal,* February 17, 1976, p. 1.
13. Atomic Industrial Forum, "Importance of Closing the Nuclear Fuel Cycle," October 1, 1976.
14. Pan Heuristics, *Moving Toward Life in a Nuclear Armed Crowd,* prepared by the Arms control and Disarmament Agency, ACDA/PAB 263, April 22, 1976, p. 5.
15. Miller, op. cit.
16. David Burnham and David Binder, "U.S. Dilemma: World Energy Need Encourages Spread of Atomic Arms," New York *Times,* October 11, 1976, p. 1.

17. Taylor and Willrich, op. cit., p. 200.

18. Energy Research and Development Administration, "U.S. Nuclear Export Activities," April 1976, Vol. 1, pp. 1–20.

19. Burnham and Binder, op. cit.

20. Energy Research and Development Administration, op. cit., Vol. 1, pp. 1–22.

21. Ann Crittenden, "Surge in Nuclear Exports Spurs Drive for Controls," New York *Times,* August 17, 1975, p. 1.

22. Murray Marden, "Curb on Atom Plant for Spain Rejected," Washington *Post,* June 22, 1976, p. 1.

23. Gerald Ford, "Statement by the President on Nuclear Policy," October 28, 1976.

24. Ibid.

25. Natural Resources Defense Council, *The Plutonium Decision* (Washington, D.C., September 1974), p. 6.

26. Atomic Energy Commission, "Special Safeguards Study" (Rosenblum Study), April 29, 1974.

27. Ibid.

28. Taylor and Willrich, op. cit., p. 20.

29. Leonard Ross, "How 'Atoms for Peace' Became Bombs for Sale," New York *Times Magazine,* December 5, 1976, p. 39.

30. Don Oberdorfer, "Taiwan to Curb A-Role," Washington *Post,* September 23, 1976, p. A-1.

31. Telephone interview with IRRC, 1974; See also Brian Jenkins, *Hearings on Proposition 15,* California State Assembly, Vol. X, November 19, 1975, p. 61.

32. Taylor and Willrich, op. cit., p. 13.

33. See John McPhee, Profiles: "Theodore B. Taylor," *New Yorker,* December 3, 10, 17, 1973; also Taylor and Willrich, op. cit., p. 25.

34. Taylor and Willrich, op. cit., p. 80.

35. Ibid., pp. 107, 108.

36. General Accounting Office, letter to Dixy Lee Ray from Henry Eschwege, October 16, 1964, p. 2.

37. Brian Jenkins, "Will Terrorists Go Nuclear?," California Seminar on Arms and Foreign Policy, November 1975, p. 3.

38. Ford, op. cit., p. 1.

39. David Lilienthal, testimony before the Senate Committee on Government Operations, January 19, 1976, p. 9.

40. Pan Heuristics, op. cit., pp. 3, 4.

41. Henry Smyth, Hearings before the House Committee on Foreign Affairs, July 9, 1974.

42. Hans Bethe, testimony before the Senate Committee on Government Operations, January 19, 1976, p. 23.

43. AEC, interview with IRRC, 1974.

44. Ross, op. cit., p. 112.

45. Pan Heuristics, op. cit., p. 19.

46. Burnham and Binder, op. cit., p. 1.

47. Adlai E. Stevenson, III, "Nuclear Reactors: America Must Act," *Foreign Affairs, October 1974, p. 68.*

48. Kathleen Teltsch, "Carter Gives Plan for Nuclear Curb," New York *Times,* May 14, 1976, p. 1.

49. Bethe, op. cit.

50. Atomic Industrial Forum, "U.S. Nuclear Export Policy," July 21, 1976.

51. Atomic Energy Commission, "Special Safeguards Study," op. cit.

52. Taylor and Willrich, op. cit.

53. Atomic Energy Commission, "Special Safeguards Study," op. cit.

54. Taylor and Willrich, op. cit.

55. Clarence Larson, "Nuclear Materials Safeguards: A Joint Industry-Government Mission," in *Proceedings of AEC Symposium on Safeguards Research and Development,* October 27–29, 1969, WASH 1147, 1969.

56. Quoted in Natural Resources Defense Council, *The Plutonium Decision,* op. cit., p. 16.

57. Tom Zito, "Report Faults Accounting of A-Materials," Washington *Post,* July 28, 1976, p. A-1.

58. Ibid.

59. Dixy Lee Ray, Hearings before the Senate Committee on Government Operations, March 13, 1974, p. 259.

60. L. Manning Muntzing, ibid, p. 308.

61. Romano Salvatori, testimony before the Michigan House of Representatives, October 24, 1974.

62. Alvin Weinberg, "Social Institutions and Nuclear Energy," *Science,* July 7, 1972, p. 27.

63. Natural Resources Defense Council, op. cit., p. 19.

64. General Atomic, "Safeguards," *Public Affairs,* 1974.

65. E. R. Johnson, *Hearings on Proposition 15,* California State Assembly, Vol. X, November 19, 1975, pp. 110, 111.

66. Taylor and Willrich, op. cit.

67. Atomic Industrial Forum "Importance of Closing the Nuclear Fuel Cycle," October 1, 1976.

68. Natural Resources Defense Council, op. cit., p. 23.

69. Russell, W. Ayres, "Policy Plutonium: The Civil Liberties Fallout," *Harvard Civil Liberties Law Review,* 1975.

70. Natural Resources Defense Council, op. cit.

71. See James Liverman, testimony before the House Committee on Interior and Insular Affairs, April 29, 1975, p. 546; Mason Willrich, "Radioactive Waste Management and Regulation," reported to ERDA, from the MIT Laboratory, September I, 1976, p. 2–6; Thomas Hollocher, "Storage and Disposal of High-Level Wastes," *The Nuclear Fuel Cycle,* Union of Concerned Scientists, October 1973, p. 9.

72. Willrich, op. cit., pp. 2–12.

73. Ibid., pp. 2–14.

74. Frank Pittman, testimony before the House Committee on Interior and Insular Affairs, April 29, 1975, p. 580.

75. Willrich, op. cit., pp. 2–13.

76. Ibid, pp. 2–14.

77. ERDA, "A National Plan for Energy Research, Development and Creating Energy Choices for the Future, 1976," Vol. 2, pp. 238 ff.

78. Ibid.

79. Ibid.

80. Lee Gapay, "Government Ponders How to Safely Dump Toxic Nuclear Waste," *Wall Street Journal,* July 26, 1976, p. 1.

81. Ibid.

82. Union of Concerned Scientists, "The Nuclear Power Issue: An Overview," n. d.

83. Hannes Alfven, speech before Ralph Nader's Critical Mass, November 18, 1975.

84. Center for Science in the Public Interest, "Nuclear Energy: The Morality of our National Policy," Washington, D.C., 1974.

85. Union of Concerned Scientists, op. cit.

86. Philip M. Boffey, "Platonium: Its Morality Questioned by the National Council of Churches," *Science,* April 23, 1976, p. 358.

87. Center for Science in the Public Interest, op. cit.
88. Hollocher, op. cit., p. 9.
89. Mason Willrich, letter to Robert Seamans, August 27, 1976.
90. Willrich, "Radioactive Waste Management and Regulation," op. cit., pp. 2–19.
91. Ibid., pp. 2–16.
92. Weinberg, op. cit.
93. F. P. Barenowski, letter to Mason Willrich, October 27, 1976, p. 8.
94. Jerry Grey, "Managing Nuclear Wastes," Atomic Industrial Forum, 1974, p. 5.
95. Quoted in Gapay, op. cit.

CHAPTER

9 ANALYSIS AND CONCLUSIONS

In attempting to analyze the impact of moral, technical, economic, and political questions on the future of nuclear power development, one is immediately struck by the enormous number of uncertainties involved. In many areas the facts are in dispute; in most areas the interpretation to be given to existing data is bitterly contested.

Persons interested in determining either whether they think continued development of nuclear power is socially responsible, or what they believe the prospects for nuclear power development to be, face the task of attempting to determine the probability that certain events will (or will not) occur. They must also determine the implications of their vision of the future. On several issues, such as whether it is appropriate to generate long-lived nuclear wastes before a permanent solution to the problem of waste disposal is found, or whether it is more responsible to deplete scarce fossil fuel resources, the final decision may be influenced by moral principles. On others, such as how effectively nuclear power can compete with coal, a more objective appraisal is possible.

This final chapter assesses, to the extent possible, the significance of the issues, the data, and the arguments of the interested parties.

DEMAND FOR ENERGY

Over the coming decades the rate at which demand for energy increases is the factor to exert the strongest influence over development of nuclear power. If growth in energy use continues at anything approaching historic rates, nuclear power will have to play a major role in the energy economy, because it seems clear that other sources of energy cannot be expanded at the rate required to meet the needs that would develop. On the other hand, if growth in demand declines dramatically, a limiting of construction of new nuclear power plants beyond those currently on order or planned could take place without unduly constraining energy supply.

We will sidestep the question of the importance of increased energy use to improved standards of living—a question that will be a central public policy issue in coming years. It does seem very likely, though, that some legislative measures will be enacted to stimulate energy conservation and efficiency. It also is apparent that the cost of generating electricity is rising rapidly and will continue to rise over the next several years. The combination of legislative measures, increased public consciousness of the possibilities for energy conservation, and rising energy costs surely will lead to a significant slowing in the rate of increase in energy use. The slowdown of growth of demand, in turn, will lessen pressures to expand development of nuclear power. Thus, it seems very unlikely that nuclear power will expand at anything close to past projections by government or industry.

Another major variable influencing the extent to which nuclear power will expand will be the rate at which production of coal increases. Large increases in production of coal are possible in the West, providing that legislation governing surface mining can be enacted, water rights arbitrated, and transportation networks established. Some optimism has been expressed recently over new discoveries of low-sulfur coal in the East. But rapid expansion of coal mining has been thwarted in the past by problems relating to mine safety, production efficiency, and environmental impact. It is not likely that these problems will be resolved in the near future.

COMPETITION WITH COAL-FIRED POWER

The second most important factor affecting the rate at which nuclear power develops probably will be how effectively nuclear power plants can compete with coal-fired plants. The costs of both nuclear and coal fuel cycles are subject to numerous uncertainties. Analysis of the available data, however, demonstrates that three factors are likely to be critical to this equation: (1) the comparative capital costs of nuclear and coal-fired power plants; (2) the comparative capacity factors at which large nuclear and coal-fired plants will operate; and (3) the prices of nuclear fuel and coal.

The capital costs of nuclear and coal plants may be substantially affected by regulatory decisions. A major component of the cost of a nuclear plant is the interest paid on funds borrowed for construction and on cost escalation during the construction phase. If the time now required between the initial application to build and the operation of a plant can be cut substantially, capital costs could be significantly reduced. For example, General Electric in its estimates of nuclear costs in 1984 assumes a nine-year construction period. If the current pace of licensing continues, however, the period will run for 11 years as it does today and add another $130 million to the cost of the plant.

Similarly, air quality standards and the cost of equipment to meet them will play an important role in the total capital costs of a coal plant. GE estimates that stack-gas scrubbers at a given coal plant will cost $80 million. Others have estimated the cost at anywhere between $50 million and $100 million.

The relative sensitivity of the costs of power generation to sulfur-dioxide regulation and to capacity factors is illustrated by a table prepared recently for Lewis J. Perl of the National Economic Research Association Inc. (See Table 2).

TABLE 2

Estimated Costs in 1990 of Energy from Nuclear Plants and Coal Plants under Alternative SO_2 Control Scenarios and at Alternative Capacity Factors (1980 dollars)

	Coal (mills per kilowatt hour)				
Capacity Factor (percent)	No Sulfur Constraints	New Source Performance Standards	Scrubbers Everywhere	Scrubbers and Low Sulfur	Nuclear
	(1)	(2)	(3)	(4)	(5)
30	47.4	52.7	53.1	57.3	56.1
40	38.4	53.7	44.1	48.3	44.2
50	32.9	38.2	38.6	43.8	37.1
60	29.3	34.6	35.0	39.2	32.3
70	26.7	32.0	32.4	36.6	28.9
80	24.8	30.1	30.5	34.7	26.4

Source: Interview with Paul Turner, June 2, 1974.

The second column assumes no restrictions on coal. The third column assumes that the plant must meet new-source performance standards of 1.2 pounds of sulfur dioxide per million BTUs. The fourth and fifth columns reflect variations—either scrubbers alone or a combination of scrubbers and low-sulfur coal—to meet federal standards to limit deterioration of air quality.

The table makes the following additional assumptions:

A 1985 cost in current dollars of $698 million for a 1,100-megawatt coal plant, $938 million for a 1,100-megawatt nuclear plant.

Scrubber costs of $100 million—2 mills per kilowatt hour.

Fuel costs in 1980 dollars for uranium of $30 per pound—7 mills per kilowatt hour; $18 per ton for high-sulfur coal—8–10 mills per kilowatt hour; $20–30 per ton for low sulfur coal—10–15 mills per kilowatt hour.
Operation and maintenance costs of 1.5 mills per kilowatt hour for both coal and nuclear plants.

Costs per kilowatt hour rise rapidly as the capacity factor declines, and because the costs for both nuclear and coal-fired power plants have increased dramatically in the last few years, plant reliabiltiy has taken on even greater importance. Commonwealth Edison has stated that its operating nuclear plants are economic at capacity factors as low as 35 percent. This may be true of plants constructed at past prices, but it appears from Perl's figures that it is less likely to be true of future plants because increased capital costs have greatly widened the spread in costs per kilowatt-hour between plants operating at 35 to 40 percent and those operating with a 70 to 80 percent capacity factor.

It is clear from a review of past performance data that industry projections comparing coal with nuclear have tended to overstate nuclear power's efficiency. Few if any plants have operated consistently at levels approaching the 70 to 80 percent capacity factor that has been used to project the costs of nuclear generation. Moreover, most comparisons of the projected costs of generating power at new nuclear and coal-fired power plants assume that both will operate at the same capacity factor. But that is not necessarily the case; one technology may prove to be more reliable than the other. And, if capacity factors differ substantially, the technology that proves more reliable will provide power at the lowest cost. For example, should new nuclear plants operate at a 50-percent capacity factor and new coal plants at a 70-percent capacity factor, the spread in capital costs would be 4.7 mills per kilowatt hour—assuming scrubbers everywhere. This would be enough to keep coal the lower-cost technology even if the price of coal rises close to $30 per ton—an event that seems unlikely. (But if nuclear plants are able to operate at a 70 percent capacity factor, they will be less expensive than any coal plant other than those that are able to operate without restrictions or emissions.)

Fuel costs, too, may affect the relative positions of coal and nuclear plants. A rapid rise in the price of uranium could affect the competitiveness of nuclear power. Uranium costs represent a relatively minor portion of total costs of generation from a nuclear plant. However, were uranium prices to double—which seems entirely possible over the next decade—the cost of generation from nuclear plants would go up to 3.7 mills. An increase of that size could make coal competitive with nuclear even if both plants were operating at a 70 percent capacity factor and the coal plants were required to use scrubbers. But the competitiveness of coal plants might also suffer from fuel price increases, from difficulties in increasing mine production, or from escalating costs of equipment required to meet environmental and safety standards.

In sum, the relative economics of coal and nuclear generation remain extremely close, both highly sensitive to a number of factors, several of which are likely to undergo changes in the future. "There are surely enough uncertainties in these numbers," Perl has concluded, "to support the views of both coal- and nuclear-fired advocates."[1]

DECISION-MAKING ISSUES

The controversy about development of nuclear power embraces institutional as well as substantive issues. Three basic conclusions can be drawn about the roles that have been played by utilities, manufacturers, the Joint Committee, the AEC, ERDA, and NRC:

First, the people involved have tended to operate on the basis of good intentions and their past experiences. There is very little evidence of bad faith or evil motives.

Second, impelled by a desire to accomplish their objectives, the supporters of nuclear power have tended to disregard opportunities to reexamine the implications of the nuclear power program and have devoted inadequate attention to the unique problems involved in nuclear power development.

Third, to some extent, particularly on the part of government agencies, institutional changes are occurring that could lead to marked shifts in the roles being played by the institutions involved.

Private Sector

Within the private sector, the electric utilities play a critical role, because they are the ones that must decide to go nuclear. For many of them the transition to nuclear power has not been easy. As described by James E. Connor, an AEC planning officer,

> The move into the nuclear age has perhaps had its greatest institutional impact on the nation's utilities. For decades, the utilities have operated as rather simple, straightforward business enterprises, ordering equipment and technical services from other firms wholly responsible for their products. Seldom was any technical capability retained by the utility itself. If a piece of equipment did not perform satisfactorily and corrective action was not taken expeditiously, the vendor could be taken to court. Utilities expected delivery of reliable, bug-free equipment and passed the safely predictable costs on to their customers. The utilities could operate in this fashion because there were few uncertainties about the costs or reliability of coal- or oil-fired plants.[2]

Some utilities, such as Commonwealth Edison, have now established large staffs of nuclear engineers. Others have relied on consultants. But many have tried to administer nuclear plants in the same way they administered fossil fuel plants. It seems clear that in the future, for utilities to succeed in using nuclear power efficiently, they will need to build up their own expertise on nuclear engineering questions.

It has been charged, with some validity, that the utilities' commitment to nuclear power—particularly in its extensiveness—was premature. Many utilities made an enormous financial commitment to nuclear power on the basis of relatively scanty performance data. But a number of events since 1973—the Arab oil embargo, the recession, rising energy costs, and lower demand—have slowed utilities' plans. A pause has occurred between the early commitment to nuclear power which will put 160 or more plants on-line by the mid-1980s, and a second generation of nuclear plants, still unordered, which would begin operation after 1985. The pause has allowed utilities time to develop and study performance data. It also has allowed them time to reexamine their plans and to consider the growing uncertainties—fuel costs, uranium availability, and recycling and waste disposal problems—that are likely to affect nuclear power in the future. As a result, utilities will not be as dependent on manufacturers' recommendations in the future as they were in the past, and it is likely that the commitment to a second generation of nuclear power will be more qualified.

As concerns the manufacturers, it seems clear that they have been and will continue to be relatively hard-nosed in their dealings with utilities and that their behavior is based on valid business considerations. GE has stated that it is reexamining its nuclear marketing strategy in order to develop a relationship with utilities and parts suppliers which would limit GE's own responsibility by having utilities share responsibility for reliability of nuclear plants. Such a decision would seem to signal that the nuclear industry has reached a certain level of maturity. It shows a recognition that nuclear plants may have problems and that manufacturers' control of reliability and projection of performance are not as complete as they once had indicated.

It is true, too, that manufacturers' representations about the safety and reliability of nuclear power often have appeared to be unduly optimistic. In part, these problems may have stemmed from an initial desire by the manufacturers to overcome public fears of nuclear power. The public image of atomic energy was strongly negative; atomic energy meant weapons. Industry has overcompensated to some degree in trying to stress safety. It has become tied to descriptions of superhuman perfection, flawless systems, and checks and counterchecks. This has made explanation of problems within the industry—many of which are minor—embarrassing, and has led the industry to be defensive where there was little need for it. The defensiveness has led to public mistrust.

A good case in point is manufacturers'—and government's—claims about plant reliability. In their efforts to stimulate nuclear development, proponents have stressed the reliability of nuclear plants. They have promised a capacity factor of 70 to 80 percent. Generally, refueling for a nuclear plant takes one month a year. Assuming yearly refueling, it would not be possible for a plant to operate at a capacity greater than 92 percent, even if it encountered no operating problems. Given the complicated nature of repairs in a radioactive environment, it appears unrealistic to leave only 12 percent leeway for unexpected down-time. Moreover, the claims of 80 percent capacity seem unnecessary. Few fossil fuel plants operate at that capacity factor, and nuclear power plants seemingly can compete effectively even if a lower capacity is attained.

Another factor affecting the flow of information from both utilities and manufacturers has been the technical nature of nuclear power. Industry has had some difficulty in deciding which audience to address. Much of industry's material is directed at others in the industry. Most of the remainder has been designed to explain nuclear technology in very simple terms to the layman. Very little has been directed at dealing seriously with those questions raised in good faith by critics.

To some extent, until 1975, industry refused to take the concerns of the critics seriously; it seems to have been more worried about what the concerns might do to the industry than about the substance of the concerns themselves. This led some persons in industry to write off critics as obstructionists, to speak of nuclear blackmail, and to concentrate more on personalities than on issues. Others, however, acknowledged the critics' questions but considered them answered. "This is a responsible and well-researched industry," one industry representative said. "We have spent a great deal of time thinking about these issues and have resolved most of them to our satisfaction. The critics are rediscovering the wheel."

The antinuclear initiatives of 1976 brought some change to the level of discussion between nuclear proponents and their critics. Many persons in industry continued to question the qualifications of persons opposed to nuclear development and to argue that their concerns were already in large part resolved. But most of them recognized the potential impact that the initiatives could have and, as a result, they began to deal with public concerns in greater depth than they had earlier. The major objective of the industry campaigns remained persuasion rather than discussion, however, and after the elections were over, three knowledgeable observers commented, "the debate over nuclear energy has become so polarized that it has stifled public consideration of options other than the two extremes of full speed ahead or a shutdown of the nuclear industry."[3]

The polarization of the debate is as much a fault of the critics of the industry as it is of the industry, and it seems fair to say, at least in some cases, that industry's reactions to the critics have been stimulated by critics' presentation of their point of view. Some critics have been careful and judicious in their pronouncements. But others have continued to use outdated studies—for example, continuing to cite the Brookhaven report as valid while criticizing Rasmussen—and to rely on discredited data. And, to a degree, many critics have shown the same unwillingness to recant that some members of industry display.

Finally, the excuses and failures of both industry spokesmen and critics seem to have been exacerbated in part by weaknesses in the mass media. Lacking sophistication on what they consider to be highly technical questions, many media representatives have simply failed to explain nuclear power issues; rather, they have only skimmed the surface, reporting the statements of industry and critics without analysis, and tending to highlight the most sensational statements without enough regard for their merit.

Public Sector

At least until recently, the Joint Committee and the AEC for the most part operated in a manner that lent credence to the arguments of the critics. They provided little opportunity for the kind of discussion of the implications of nuclear power that opponents seek, and they were inclined to ignore skeptics and get on with the job of development. In addition, because of their recognized expertise in this complicated area, persons in Congress and the Executive Branch tended to follow their lead, rather than asking questions of their own or examining the questions that critics were raising.

The situation in the AEC changed somewhat in the early 1970s. The regulatory staff expanded greatly, and the agency began to make much more information available to the public. The greater openness of the AEC after 1972 was carried over into the NRC, and there is no doubt that the amount of information now available to the public is far more extensive than was once the case.

Nevertheless, concerns about the NRC's willingness to provide the public with information on problems in the nuclear program persist. Resignations by two NRC engineers, an open letter by an NRC consultant, and newspaper articles describing NRC officials unhappy with the agency's handling of internal dissent have contributed to these concerns. There is little reason to believe that either the concerns or the complaints about the NRC's ability or willingness to deal with them publicly will be resolved in the near future. It will continue to be difficult for the NRC to release enough information to satisfy critics or to overcome the preception that the agency is not sufficiently open.

Changes within Congress have greatly facilitated discussion of nuclear issues, and now that nuclear power has reached the national political forum, it very likely will continue to be a major subject of congressional inquiry.

Exclusive oversight responsibility has slipped away from the Joint Committee on Atomic Energy to a number of committees and subcommittees. This process is continuing as plans are made for a final dismembering of the Joint Committee, and as members of Congress, spurred by constituents' concern as a result of the antiinitiative campaigns, take a more active interest in nuclear power as a political issue. Critics appear primed to take advantage of changes in Congress to pursue at a visible level what they see as the need for further debate over the future of nuclear power, and they are likely to focus on a number of bills—on subjects including export of nuclear materials, government assistance to enrichment or recycling, and the ERDA budget—in the coming years as vehicles for debate.

At the same time, proposed legislation could help the industry. Efforts to speed the licensing process could lead to changes beneficial to nuclear power's growth. Reduced time spent in the licensing process would cut overall time between beginning construction and operation, which—with high interest rates and rapidly escalating costs for labor and materials—has been an important factor in the high capital cost of a nuclear plant. And decisions on export policy and aspects of the fuel cycle—enrichment, recycling, and waste disposal—would clarify both opportunities for and limitations on the industry's development.

SAFETY CONCERNS

The debate over safety that critics of nuclear power are trying to stimulate focuses primarily on concerns about dangers associated with nuclear power plants and the nuclear fuel cycle. The major issue as posed by opponents of nuclear power is whether the electric utilities, with the support of industry and government, should make a commitment to a form of technology that may pose unprecedented threats not only to the present generation but to countless future generations. The threats arise from the highly radioactive character of the materials in the nuclear fuel cycle, which continues for extremely long periods.

Alvin Weinberg, former director of the AEC's Oak Ridge National Laboratory, described the issue in a 1972 article in *Science*: "When nuclear energy was small and experimental and unimportant, the intricate moral and institutional demands of a full commitment to it could be ignored or not taken seriously. Now that nuclear energy is on the verge of becoming our dominant form of energy, such questions as the adequacy of human institutions to deal with this marvelous new kind of fire must be asked and answered, soberly and responsibly." The commitment to nuclear energy Weinberg described as a "Faustian bargain": "On the one hand, we offer—in the catalytic nuclear burner—an inexhaustible source of energy. . . . But the price we demand of society for this magical energy source is both a vigilance and a longevity of our social institutions that we are quite unaccustomed to." Weinberg said the issues to be resolved are "transcientific," to be "adjudicated by a legal or political process rather than by scientific exchange."[4] He concluded that the benefits of developing nuclear power are greater than any risks involved. Others, such as the Center for Science in the Public Interest, are more concerned by what they see as the "irreversible consequences" of nuclear power development, and urge that it be halted.[5]

The concerns of opponents of nuclear power are likely to have the greatest impact on industry if they are picked up by political leaders and incorporated in new legislation. Predicting whether and in what forms such a chain of events is likely to occur is very difficult. It seems evident, though, that the increasing attention various activist groups are devoting to nuclear power issues has succeeced in giving them greater public visibility, and that increased consideration of these issues in political forums has resulted. Also evident is the prospect that a single major untoward event involving nuclear power—a serious accident, for example, or the hijacking of a shipment of plutonium—is likely to lead to swift and dramatic political action of a kind that could cripple development of the industry.

Investors' evaluations of these issues also could have a significant impact on the industry. As this book mentioned earlier, utilities will need to rely heavily on external financing in order to build new nuclear power plants. Should a substantial number of institutional investors decide, for either moral or economic reasons, to curtail their financial involvement in nuclear power, the result would be a severe constraint on nuclear power's expansion.

Nuclear Power Safety

The dispute over the safety of nuclear power is one of the most troublesome to resolve. The conflicting interpretations of past performance data, in particular, present observers with a classic problem of which Ph.D. to believe. Both sides argue that the data support their position. Implications to be drawn from the data are still somewhat unclear.

From a safety point of view, nuclear plant operations have been successful. There have been no fatal accidents. There have been some new and some recurrent problems, but many of these abnormal occurrences were anticipated, and back-up systems have prevented them from becoming serious problems. Others were not anticipated, but have also been caught by monitoring or safety systems before they caused unacceptable damages. Plant design has been adequately conservative to prevent accidents with serious consequences.

What this record implies for the future is not so clear. Experience with some 61 reactors operating for fewer than 20 years may be very different from that with hundreds of larger reactors operating over a 50-year span. Standardized plant design and more experience may cut down on the number of abnormal occurrences. Larger plants and new technologies might, on the other hand, increase problems.

Disagreements over the adequacy of requirements for emergency core-cooling systems were partially resolved by the AEC's revision of the relevant regulations. But a permanent resolution of the controversy is unlikely in the near future. In its report on reactor safety, the American Physical Society stated that although it had no reason to doubt that the ECCS would work, no quantitative basis existed for evaluating ECCS performance. The society doubted that a complete quantitative evaluation could ever be done through the current program.[6] At least some critics say that they will not be fully satisfied until the NRC has conducted loss-of-fluid tests on power reactors considerably larger than the one on which tests have been scheduled. For the nonexpert observer, reaching a confident judgment on this issue is extremely difficult.

Similarly, drawing conclusions about the weight to be given to the Rasmussen report presents major difficulties. The report, as described by the Environmental Protection Agency, was "an innovative forward step in risk assessment of nuclear reactors,"[7] but it does not represent the final word in safety assessment. As the draft report itself said, "the use of quantitative techniques in decision-making associated with risk is still in its early stages and is highly formative."[8] EPA, the Union of Concerned Scientists, and the Sierra Club also argue that the study contains errors and is misleading in some of its conclusions.

Critical questions about the report—the answers to which again depend upon sophisticated technical judgments—are how well the Rasmussen team assessed the risks that events or faults would occur, and to what extent the techniques they used improved upon the similar techniques that were used, with not entirely satisfactory results, in the aerospace program. In short, even if the report is the best assessment possible, is it good enough to serve as a basis for the decision to develop nuclear power? The Advisory Committee on Reactor Safeguards concluded that "a continuing effort and better data will be required to evaluate the validity of the quantitative results [of the Rasmussen study] on absolute terms."[9]

Safeguards: Issues

On questions involving safeguards, many of the predictions required relate more to human and national behavior than to technology. Agreement exists on the need to safeguard plutonium and other special nuclear materials.

But the consensus breaks down when it comes to assessing how great the risks are and what particular policies should be implemented to guard against them.

The risk that appears the greatest is the prospect that other countries will reprocess reactor wastes to acquire plutonium for construction of nuclear weapons. India has demonstrated that diversion of wastes is possible. The International Atomic Energy Agency cannot prevent diversion; it only has authority to keep track of special nuclear materials, and some observers question how effectively it does the job. "If a country is determined to acquire nuclear weapons, there are no practical technical controls on nuclear power programs to prevent it from doing so," Theodore Taylor and Harold Feiverson commented in the *Bulletin of the Atomic Scientists,*[10] and active export programs that include reprocessing and enrichment technology can only serve to make development of an independent nuclear-weapons capacity easier. Thus, it seems probable that the international development of nuclear power will facilitate proliferation of nuclear weapons—a trend that would be inconsistent with the U.S. government's policy.

The United States has come to recognize this likelihood and for the last year has been exploring methods for limiting it. U.S. representatives have discussed bilateral arrangements that would assure the United States full control over the fuel cycle and would provide for multinational fuel centers with international controls. Most recently, it has decided, unilaterally, not to export either reprocessing or enrichment technology.

The U.S. efforts have won support from two other major exporters of nuclear technology—France and West Germany—both of which have agreed to reexamine their own export policies. A halt to the export of more sophisticated nuclear facilities may have two effects. The exporting countries may provide assurances of a permanent supply of fuel to countries with nuclear plants operating or planned. In this case, there should be little impact on the sales of reactors themselves. The exporting countries may find themselves unable to provide an assured supply of fuel. In this case, either the international demand for reactors will fall or countries will try to develop their own reprocessing and enrichment capacities.

In order to minimize the potential of proliferation, therefore, the exporting countries either must agree to become permanent suppliers of fuel or must persuade other countries to curtail their development of nuclear power. For the moment, the United States appears to have taken the first course. It is seeking to enlarge its own enrichment capacity, but refusing to export enrichment technology; it will probably reach at least a temporary decision on reprocessing in the next two years, but it will not export reprocessing equipment. The alternative would be to halt nuclear development at home and to halt exports of reactors at the same time. Any lesser action would be insufficient to prevent other countries with active nuclear power programs from supplying what is now a U.S.-dominated market. It is not clear that a U.S.

moratorium would stop the trade in nuclear power plants, because several other countries have made major commitments to nuclear power for political or economic reasons. But a halt in the U.S. program would be such a dramatic development, it seems clear, that it would stimulate at least a serious reconsideration of the subject on the international level.

The threat that terrorist groups—foreign or domestic—will obtain plutonium and construct nuclear weapons seems lesser, but by no means nonexistent. One line of defense would be to require more stringent protection measures for nuclear facilities and radioactive materials. A second would be to abandon plans for recycling plutonium and for development of breeder reactors.

The principal argument in support of the breeder and recycling is that supplies of uranium are limited. The NRC and industry representatives have stated that domestic uranium production capacity might not be adequate to meet domestic requirements by the mid-1980s. Many observers, however, do not agree that the United States is about to run short of uranium, and there appears to be some factual support for their disagreement. Milton Searl, in a recent study for the Electric Power Research Institute, notes that "a number of geologists regard the state of uranium exploration as being roughly comparable to the days when oil was found on the basis of oil seeps." He says "there is even reason to believe that the nation may not yet have found its best deposits of uranium."[11] An official of Kerr-McGee Corporation told IRRC in 1974, "There are large areas prospective for uranium that have had little or no exploration. . . . While the history of uranium is short, the industry has shown the capacity to supply the market when given sufficient lead time and adequate economic incentives."[12]

The prospects for development of the breeder appear to be diminishing, as a consequence of both economic and political concerns. There is considerable doubt over the economics of reprocessing and controversy over the cost of the breeder itself. One industry official remarked to IRRC that as a result of the reorganization of the AEC, "the breeder program has become increasingly vulnerable."

Waste Disposal: Issues

In many ways, the questions raised about disposing of nuclear power plant wastes parallel those raised about safeguards. The greatest uncertainty lies in predicting human behavior over an extended period of time.

It seems unlikely that concerns about waste disposal and ERDA's failure to develop a long-term solution to the waste disposal problem will lead in and of themselves to a political decision to halt nuclear power development. On the other hand, debate over the question is causing delays in completion of

ERDA's interim storage facility, which in turn could impair the operations of the industry. Until ERDA decides upon an acceptable method of disposal, nuclear critics will continue to argue against creation of new waste.

CONCLUSION

Economic, political, and technological uncertainties continue to plague the nuclear industry. A major commitment to a second generation of light-water reactors is not likely to come soon, or at least anywhere near as soon as once was predicted. Recycling of plutonium will undergo careful scrutiny and may well be postponed indefinitely until an economic justification for its development becomes more apparent. Without reprocessing, the breeder program will face even greater skepticism, and while it may not be abandoned, its development will suffer new delays.

Assessments of nuclear power's potential role in meeting energy demand in the future have changed dramatically in the last two years. What was viewed two years ago as the primary energy resource is viewed increasingly as a resource of last resort. Increasingly, energy planners as well as politicians are emphasizing energy conservation before growth, and coal before nuclear. Nuclear development will be limited to that which is necessary, and efforts are under way to limit that necessity. In his speech to the nation in February 1977, President Carter stated: "Our program will emphasize conservation. The amount of energy now being wasted which could be saved is greater than the total energy we are importing from foreign countries. We will also stress development of our rich coal reserves in an environmentally sound way; we'll emphasize research on solar energy and other renewable energy sources; and we'll maintain strict security on necessary atomic energy production."

NOTES

1. Lewis J. Perl, "Is Nuclear Energy Economically Effective," *Energy Magazine,* Summer/-Fall 1976, p. 23.
2. James E. Connor, "Prospects for Nuclear Power," reprinted from "The National Energy Problem," *Proceedings of the Academy of Political Science,* December 31, 1973, p. 67.
3. Harold A. Feiverson, Theodore B. Taylor, Frank von Hippel, and Robert H. Williams, "The Plutonium Economy," *Bulletin of the Atomic Scientists,* December 1976, p. 17.
4. Alvin Weinberg, "Social Institutions and Nuclear Energy," *Science,* July 7, 1972, p. 33.
5. Center for Science in the Public Interest, "Nuclear Energy: The Morality of Our National Policy" (Washington, D.C., 1974).
6. *Report to the American Physical Society by the Study Group on Light Water Reactor Safety, 28 April 1975,* National Science Foundation and the Atomic Energy Commission, pp. I–5, 6.
7. Environmental Protection Agency, "Comments on the Reactor Safety Study," November 1974.

8. Atomic Energy Commission, "An Assessment of Accident Risks in U.S. Commercial Nuclear Power Plants," Wash 1400, August 1974.

9. Advisory Committee on Reactor Safeguards, quoted by David Okrent, *Hearings on Proposition 15,* California State Assembly, Vol. VI, October 29, 1975, p. 87.

10. Harold A. Feiverson, and Theodore B. Taylor, "Security Implications of Alternative Fission Futures," *Bulletin of the Atomic Scientists,* December 1976, p. 17.

11. Electric Power Research Institute, *Uranium Resources to Meet Long-Term Uranium Requirements,* September 4, 1974.

12. Telephone interview with IRRC, 1974.

ABOUT THE AUTHOR

DESAIX MYERS III, Associate Director and Senior Research Analyst with the Investor Responsibility Research Center in Washington, D.C., has written a number of articles relating to public policy and corporate social responsibility. A major focus of his work has been on environmental and energy issues.

Before he joined IRRC, Mr. Myers was a desk officer with the Agency for International Development in Washington, D.C. He also served as a field officer with AID in East Pakistan (Bangladesh) during the years 1970 and 1971.

Mr. Myers received his Master's degree in international development studies from the Fletcher School of Law and Diplomacy in 1969, writing his dissertation on ideology and rural development in Kenya. He is the author of several reports on coal strip mining, emissions from coal-fired power plants, and nuclear power. He has also written articles on rural development, he has contributed a chapter to *Disaster in Bangladesh,* and written *The Labor Practices of U.S. Corporations in South Africa,* published by Praeger Publishers.

RELATED TITLES

Published by
Praeger Special Studies

ALTERNATIVE ENERGY STRATEGIES: Constraints and Opportunities
John Hagel III

THE ECONOMICS OF NUCLEAR AND COAL POWER
Saunders Miller

THE ENERGY CRISIS AND THE ENVIRONMENT: An International Perspective
Donald R. Kelley

*THE ENERGY CRISIS AND U.S. FOREIGN POLICY
edited by Joseph S. Szyliowicz
and Bard E. O'Neill

*THE UNITED STATES AND INTERNATIONAL OIL: A Report for the Federal Energy Administration on U.S. Firms and Government Policy
Robert B. Krueger

*Also available in paperback as a PSS Student Edition.